STUDENT SOLUTIONS MANUAL

EIGHTH EDITION

STATISTICS

A FIRST COURSE

JOHN E. FREUND · BENJAMIN M. PERLES

PEARSON

Prentice
Hall

Upper Saddle River, NJ 07458

Editor-in-Chief: Sally Yagan
Acquisitions Editor: Petra Recter
Supplement Editor: Joanne Wendelken
Assistant Managing Editor: John Matthews
Production Editor: Allyson Kloss
Supplement Cover Manager: Paul Gourhan
Supplement Cover Designer: Joanne Alexandris
Manufacturing Buyer: Ilene Kahn

© 2004 Pearson Education, Inc.
Pearson Prentice Hall
Pearson Education, Inc.
Upper Saddle River, NJ 07458

Pearson Prentice Hall® is a trademark of Pearson Education, Inc.

The author and publisher of this book have used their best efforts in preparing this book. These efforts include the development, research, and testing of the theories and programs to determine their effectiveness. The author and publisher make no warranty of any kind, expressed or implied, with regard to these programs or the documentation contained in this book. The author and publisher shall not be liable in any event for incidental or consequential damages in connection with, or arising out of, the furnishing, performance, or use of these programs.

Printed in the United States of America

ISBN 0-13-046655-7

Pearson Education Ltd., *London*
Pearson Education Australia Pty. Ltd., *Sydney*
Pearson Education Singapore, Pte. Ltd.
Pearson Education North Asia Ltd., *Hong Kong*
Pearson Education Canada, Inc., *Toronto*
Pearson Educación de Mexico, S.A. de C.V.
Pearson Education—Japan, *Tokyo*
Pearson Education Malaysia, Pte. Ltd.

Table of Contents

Chapter

Chapter 1

Introduction

Note: Answers that are given in the Solutions to Practice Exercises at the end of each chapter of the text are not repeated here.

1.1   Practice exercise.

1.3   Single, 9; Married, 7; widow or widower, 3; divorced, 5.

1.5   (a)   The data is viewed as a population if we are interested in the tips received by waitresses employed in this restaurant during this week
          (b)   The data is viewed as a sample if we are studying waitresses in different restaurants or in different time periods.

1.7   Practice exercise.

1.9   (a)   If the filling machine was defective and failed to fill (for example) alternate cans completely, then the inspector would see only the cans that were filled correctly, or only the cans that were not filled correctly. He would not see a random sample of correctly and incorrectly filled cans.
          (b)   Randomly selected telephone directory pages are often not representative of the population of the city. For example, if the page is filled with names beginning with Mac... or Mc..., their ethnic backgrounds might differ from those of Hispanic or other backgrounds. Or if the page is opened to the names Saint or St. the telephone calls would reach a disproportionate number of schools, hospitals and churches with differing views on some subjects than the general population of the city. In any event, the use of the telephone to conduct interviews risks missing a disproportionate number of low income people who cannot afford telephone service, but nevertheless go to the polls and vote.
          (c)   Some products, including grains, settle in moving freight cars. The heaviest portion moves to the bottom of the car and the lightest remains on top. To sample the product we must also take small quantities from the sides, bottom, and throughout the car.

1.11 (a)   Descriptive. Numbers can be read directly from the data.
          (b)   Descriptive. Numbers can be calculated from the data, $21 - 19 = 2$ thousand
          (c)   Descriptive. Numbers can be calculated from the data, $34 - 26 = 8$ thousand
          (d)   Generalization. There may be many other reasons for the increase in unemployment.

1.13 Practice exercise.

1.15 (a)   Descriptive. This conclusion can be read from line 1 of the table.
          (b)   Generalization. This conclusion cannot be determined from the data.
          (c)   Descriptive. This conclusion can be obtained by reading columns 2 and 3 of the table.
          (d)   Descriptive . This conclusion can be obtained by reading columns 2 and 3 of the table.
          (e)   Generalization. This conclusion cannot be determined from the data.
          (f)   Generalization. This conclusion cannot be determined from the data.

Chapter 2

Summarizing Data: Listing and Grouping

2.1   Practice exercise.

2.3

```
                    •
             •      •
             •      •
             •      •
             •      •
             •      •
             •      •
             •      •          •
             •      •          •
      •      •      •          •
      0      1      2          3
```

2.5   Coffee   • • • • • • • • • • • • • • • •
            Soda pop   • • • • • • • • • •
            Juice   • • • • • • •
            Tea   • • • • •
            None   • • • •

2.7   16 | 8 4
            17 | 6 1 3
            18 | 9 4 8 1
            19 | 5 3 9 8 1 7 5 7
            20 | 7 3 0 2
            21 | 6 4
            22 | 7

2.9   (a)   10, 11, 11, 12, 15, 17, 18;
            (b)   120, 122, 123, 123, 125;
            (c)   0.60, 0.62, 0.63, 0.66;
            (d)   1.50, 1.51, 1.51, 1.53, 1.54. 1.56.

2.11

5*	4 2
5.	6 9 7 6
6*	2 2 0 0 1 3 4 0
6.	6 7 8 7 5 9 5 8
7*	1 3 1
7.	8 6
8*	0

2.13

2.	9 8
3*	3 1 1 1 4 0
3.	5 7 6 8 9 8 7 6
4*	3 3 0 1 0 4 4
4.	6 6 9 7
5*	1 0
5.	6

2.15   Practice exercise.

2.17   One possibility is 0–249,999; 250,000–499,999; 500,000–749,999; 750,000–999,999; 1,000,000–1,249,999; 1,250,000–1,499,999; 1,500,000–1,749,999; 1,750,000–1,999,999.

2.19   (a) No;
        (b) Yes;
        (c) No;
        (d) Yes.

2.21   Practice exercise.

2.23   (a) 0, 20, 40, 60, 80, 100, 120, and 140;
        (b) 19, 39, 59, 79, 99, 119, 139, and 159;
        (c) 9.5, 29.5, 49.5, 69.5, 89.5, 109.5, 129.5, and 149.5;
        (d) 20.

2.25   Practice exercise.

2.27   (a) 31.5, 40.5, 49.5, 58.5, 67.5, 76.5, 85.5, 94.5.
        (b) 32–40, 41–49, 50–58, 59–67, 68–76, 77–85, and 86–94.

2.29   There is no provision for 11 and 31. Also, there is ambiguity because 23 can be put into the fourth class or the fifth class.

2.31 (a)

Justices of the U.S. Supreme Court	
Years of Service	Percentage
0–4	9
5–9	25
10–14	15
15–19	25
20–24	11
25–29	7
30–34	5
35–39	2
Total	100 (Approx)

(b)

Justices of the U.S. Supreme Court	
Years of Service	Percentage
Less than 5	9
Less than 10	34
Less than 15	49
Less than 20	74
Less than 25	85
Less than 30	92
Less than 35	98
Less than 40	100

(c)

Justices of the U.S. Supreme Court	
Years of Service	Percentage
More than 0	100
More than 5	91
More than 10	66
More than 15	51
More than 20	26
More than 25	15
More than 30	8
More than 35	2
More than 40	0

2.33 (a)

Age	Cumulative Frequency
Less than or equal to 19	0
Less than or equal to 24	129
Less than or equal to 29	350
Less than or equal to 34	660
Less than or equal to 39	823
Less than or equal to 44	928
Less than or equal to 49	990
Less than or equal to 54	1,000

(b)

Age	Cumulative Frequency
20 or more	1,000
25 or more	871
30 or more	650
35 or more	340
40 or more	177
45 or more	72
50 or more	10
55 or more	0

2.35   In 7, 8, 8, and 17 World Series, the defeated team won 0, 1, 2, and 3 games, respectively.

2.37   The frequencies corresponding to excellent, very good, good, fair, poor, and very poor, are 3, 9, 20, 6, 1, and 1, respectively.

2.39

Dollars	Cumulative f
Less than 1	0
Less than 21	18
Less than 41	80
Less than 61	143
Less than 81	186
Less than 101	200

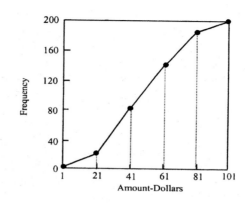

2.41

Thousands of Dollars	Cumulative Frequency
Less than 50	0
Less than 75	3
Less than 100	24
Less than 125	43
Less than 150	46
Less than 175	47
Less than 200	50

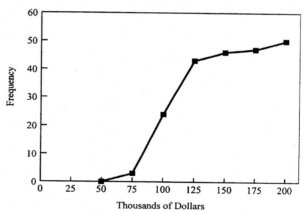

2.43   The central angles are 16.7°, 230.5°, 53.2°, 37.4°, 19.1°, 13.2°.

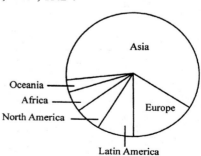

2.45   The central angles for the degrees are Bachelor's 192°; Doctor's, 8°; Master's, 70°; Associate's 90°.

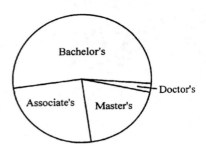

Chapter 3

Summarizing Data: Statistical Descriptions

3.1   Practice exercise.

3.3   (a) If the research department is interested only in the effect of the pill on the voluntary subjects but will not use the information for other purposes.
(b) If the pill is effective on the voluntary subjects, the research department will assume that it will have the same effect on the general population.

3.5   a. $75.3 + 79.9 + 78.0 + 63.2 + 79.6 + 75.9 + 71.6 + 63.7 + 78.9 + 77.6 + 78.6 + 66.1 + 80.2 + 81.7 + 74.1 + 79.8 + 68.8 + 80.2 + 66.2 + 77.8 + 77.3 = 1,574.5$

$$\frac{1,574.5}{21} \approx 75.0 \text{ years.}$$

b.   To get a more accurate expectation of life at birth, we should weight each nation's expectation by the size of its population.

3.7 $\dfrac{511 + 493 + 491 + 744 + 664 + 789 + 615 + 821}{8} = \dfrac{5,128}{8} = \$641$

3.9 $\dfrac{4 + 3 + 0 + 0 + 2 + 4 + 4 + 3 + 1 + 0 + 2 + 0 + 2 + 1 + 4}{15} = 2$

The mean is 2, but as used here, the term "average criminal" is too vague.

3.11 $168 \times 64 = 10,752$, which is less than the maximum load of 12,000 pounds.

3.13 $\bar{x} = 7.3$ hours. It is useful but it conceals the variation of the numbers around their average.

3.15 $\bar{x} = $ a comfortable $85°$, but the mean conceals the variation in the temperatures which include some very hot days.

3.17   Practice exercise.

3.19 $\dfrac{(300)(19.0) + (1,340)(21.2) + (534)(17.8)}{300 + 1,340 + 534} = \dfrac{43,613.2}{2,174} \approx 20.1$ thousands of pounds.

3.21 $\dfrac{(488)(0.307) + (137)(0.299) + (646)(0.297) + (533)(0.291) + (502)(0.283)}{488 + 137 + 646 + 533 + 502} = \dfrac{679.81}{2,306} \approx 0.295$

3.23 $\dfrac{(12)(5,000) + (24)(6,200) + (14)(6,900)}{50} = \dfrac{305,400}{50} = \$6,108$

3.25   Practice exercise.

3.27   (a) $\frac{39+1}{2} = 20$, so that the median is the 20th value;

        (b) $\frac{150+1}{2} = 75.5$, so that the median is the mean of the 75th and 76th values.

3.29   Arranged according to size, the data is 0, 1, 1, 3, 5, 5, 5, 6, 6, 8, 10, 12, and 15. Since $\frac{13+1}{2} = 7$, the median is the 7th value, which is 5 meals.

3.31   Arranged according to size, the data is

12	18	26	28	31	33	40	44	45	49	53	58
61	63	75	80	80	89	96	103	125	125	127	129

$\frac{24+1}{2} = 12.5$, so that the median is the mean of the 12th and 13th values, namely, $\frac{58+61}{2} =$ 59.5 power failures.

3.33 $\frac{400+1}{2} = 200.5$, so that the median is the mean of the 200th and the 201st values. Since there are 72 zeros, 106 ones, and 153 twos, the 200th and 201st values are both twos and the median is $\frac{2+2}{2} = 2$.

3.35   Since $\frac{20+1}{2} = 10.5$, the median is the mean of the 10th and 11th values. There are three values on the first stem and eight on the second stem, so that the 10th and 11th values are the two largest values on the second stem, and the median is $\frac{68+69}{2} = 68.5$.

3.37   Practice exercise.

3.39   Experiments will produce various results.

3.41   Since the three midranges are 29.8, 30.0, and 30.3, the manufacturers of car C can use the midrange to substantiate the claim that their car performed best in the test.

3.43   Since $\frac{18+1}{2} = 9.5$, the median is the mean of the 9th and 10th values. Since $\frac{9+1}{2} = 5$, Q_1 is the 5th value and Q_3 is the 5th value counting from the other end, namely, the 14th value counting from left to right.

3.45   Practice exercise.

3.47

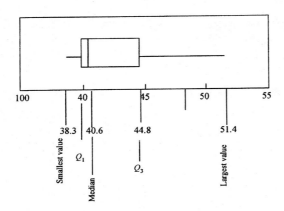

3.49   (a)   The mean is $\dfrac{4+3+3+5+...+3}{15} = \dfrac{65}{15} = 4\frac{1}{3}$ eggs.

      (b)   Arranging the data according to size, we get 2, 3, 3, 3, 4, 4, 4, 4, 4, 4, 5, 5, 5, 7, and 8. The location of the median is $\dfrac{15+1}{2} = \dfrac{16}{2} = 8$ and counting from either end of the array the 8th value is the median, which is 4 eggs.

      (c)   The location of quartile 1 is $\dfrac{7+1}{2} = \dfrac{8}{2} = 4$, and counting from the lowest value in the array we find that Q_1 is 3 eggs. The location of Q_3 is obtained by again counting 4 items, beginning with the largest, and we find that Q_3 is 5 eggs.

      (d)   The midquartile is $\dfrac{Q_1 + Q_3}{2} = \dfrac{3+5}{2} = \dfrac{8}{2} = 4$ eggs.

      (e)   The midrange is the mean of the smallest and largest values, which is $\dfrac{2+8}{2} = \dfrac{10}{2} = 5$ eggs.

3.51   (a)   The mode is 4 percent since there are more fours than any other value.

      (b)   Arranging the data according to size, we get

$$2\ \ 2\ \ 3\ \ 3\ \ 3\ \ 4\ \ 4\ \ 4\ \ 4\ \ 4\ \ 4\ \ 4\ \ 4\ \ 4\ \ 5\ \ 6\ \ 6$$

           Since the location of the median is $\dfrac{18+1}{2} = 9.5$, its value is $\dfrac{4+4}{2} = 4$ percent.

      (c)   To obtain the mean, sum the 18 values and divide by 18. Thus, the mean equals

$$\frac{4+3+3+3+4+2+5+4+4+4+4+6+2+6+4+4+4+4}{18}$$

           ≈ 3.9 percent.

3.53   Since 0, 3, 8, 10, 12, and 15 each occur once, 1 and 6 each occur twice, and 5 occurs three times, the mode is 5.

3.55   Practice exercise.

3.57 There were 13 purchases with credit card, 12 with store charge, 7 with check, and 6 with cash. The mode is 13 purchases with credit card.

3.59 Practice exercise.

3.61 For stock A, the range is $18.41 - 17.03 = \$1.38$.
For stock B, the range is $20.60 - 19.55 = \$1.05$.
Thus, stock B is less variable than stock A.

3.63 The range is $2.93 - 0.09 = 2.84$ kilograms per cubic meter.

3.65 (a)

x	$(x - \bar{x})$	$(x - \bar{x})^2$
38	−7.7	59.29
45	−0.7	0.49
53	+7.3	53.29
44	−1.7	2.89
49	+3.3	10.89
45	−0.7	0.49
47	+1.3	1.69
48	+2.3	5.29
46	+0.3	0.09
42	−3.7	13.69
457		148.1

$$\bar{x} = \frac{457}{10} = 45.7 \qquad s = \sqrt{\frac{148.1}{10-1}} = \sqrt{16.444} \approx 4.06$$

(b) $\sum x = 457, \sum x^2 = 21{,}033, S_{xx} = 21{,}033 - \frac{(457)^2}{10} = 148.1$

and $s = \sqrt{\frac{148.1}{9}} \approx 4.06$

3.67

x	x^2
3.7	13.69
2.2	4.84
11.9	141.61
3.7	13.69
8.2	67.24
5.1	26.01
0.8	0.64
1.2	1.44
36.8	269.16

$\sum x = 36.8, \sum x^2 = 269.16, n = 8$, so

$S_{xx} = 269.16 - \frac{(36.8)^2}{8} = 99.88,$

and $s = \sqrt{\frac{99.88}{8-1}} \approx \$3.8$ thousand

3.69

x	x^2
21	441
30	900
43	1,849
66	4,356
31	961
78	6,084
44	1,936
30	900
343	17,427

$\sum x = 343,\ \sum x^2 = 17{,}427,\ n = 8,$ so

$S_{xx} = 17{,}427 - \dfrac{(343)^2}{8} = 2{,}720.875,$

and $s^2 = \dfrac{2{,}720.875}{7} \approx 388.7$ vetoes

and $s = \sqrt{388.7} \approx 19.7$ vetoes

3.71 $\sigma = \sqrt{\dfrac{S_{xx}}{N}}$ where $S_{xx} = \sum x^2 - \dfrac{\left(\sum x\right)^2}{N}$

x	x^2
31.3	979.69
10.6	112.36
30.0	900.00
42.0	1,764.00
44.0	1,936.00
37.0	1,369.00
73.5	5,402.25
268.4	12,463.30

$\sum x = 268.4,\ \sum x^2 = 12{,}463.30,\ N = 7,$ so

$S_{xx} = 12{,}463.30 - \dfrac{(268.40)^2}{7} \approx 2{,}172.08,$

and $\sigma = \sqrt{\dfrac{2{,}172.08}{7}} \approx 17.62$ billions of kWh.

3.73 $\sum x = 21, \sum x^2 = 123,\ n = 4, S_{xx} = 12.75$, and $s = \sqrt{\dfrac{12.75}{4-1}} \approx 2.06.$ The range is $7 - 3 = 4$. This supports the claim that for samples of size $n = 4$ the range should be roughly twice as large as the standard deviation.

3.77   Practice exercise.

3.79   For nearly all sets of data, the actual percentage of data lying between the limits is much greater using the empirical rule than with Chebyshev's theorem.

3.81   (a)   The percentage is at least $\left(1 - \dfrac{1}{5^2}\right) \cdot 100\% = \dfrac{24}{25} \cdot 100\% = 96\%.$

      (b)   The percentage is at least $\left(1 - \dfrac{1}{8^2}\right) \cdot 100\% = \dfrac{63}{64} \cdot 100\% \approx 98.44\%.$

      (c)   The percentage is at least $\left(1 - \dfrac{1}{10^2}\right) \cdot 100\% = \dfrac{99}{100} \cdot 100\% = 99\%.$

      (d)   The percentage is at least $\left(1 - \dfrac{1}{20^2}\right) \cdot 100\% = \dfrac{399}{400} \cdot 100\% = 99.75\%.$

3.83   (a)   About 68% of the flights will arrive within this interval.
         (b)   About 95% of the flights will arrive within this interval.

(c)　About 99.7% of the flights will arrive within this interval.

3.85　(a)　Since $1 - \dfrac{1}{k^2} = 1 - \dfrac{1}{2^2} = \dfrac{3}{4}$, we get $k = 2$; $45 \pm (2)(6) = 45 \pm 12$, and we find that at least $\dfrac{3}{4}$ of the values must fall between 33 and 57.

　　　(b)　Since $1 - k^2 = 1 - \dfrac{1}{4^2} = \dfrac{15}{16}$, we get $k = 4$; $45 \pm (4)(6) = 45 \pm 24$, and we find that at least $\dfrac{15}{16}$ of the values must fall between 21 and 69.

3.87　(a)　Since $z = \dfrac{26 - 21.4}{3.1} \approx 1.48$ and $z = \dfrac{26 - 22.1}{2.8} \approx 1.39$ for the two universities, the student is in a relatively better position with respect to the first university.

　　　(b)　Since $z = \dfrac{31 - 21.4}{3.1} \approx 3.10$ and $z = \dfrac{31 - 22.1}{2.8} \approx 3.18$ for the two universities, the student is in a relatively better position with respect to the second university.

3.89

Age In Years		Weight in Pounds	
x	x^2	x	x^2
22	484	115	13,225
18	324	159	25,281
26	676	141	19,881
20	400	137	18,769
24	576	130	16,900
110	2,460	682	94,056

(a)　Mean age is $\dfrac{110}{5} = 22$ years

　　　S_{xx} for age is $2{,}460 - \dfrac{(110)^2}{5} = 40$, and $s = \sqrt{\dfrac{40}{5-1}} \approx 3.16$ years

(b)　Mean weight is $\dfrac{682}{5} = 136.4$ pounds

　　　S_{xx} for weight is $94{,}056 - \dfrac{(682)^2}{5} = 1{,}031.2$ and $s = \sqrt{\dfrac{1{,}031.2}{5-1}} \approx 16.06$ pounds

(c)　V for age is $\dfrac{3.16}{22} \cdot 100 \approx 14.4\%$. V for weight is $\dfrac{16.06}{136.4} \cdot 100 \approx 11.8\%$.

　　　Since 14.4 is greater than 11.8 the age data is relatively more variable than the weight data.

3.91　Practice exercise.

3.93

Degrees Fahrenheit	x	f	xf	x^2f
100 – 104	102	2	204	20,808
105 – 109	107	9	963	103,041
110 – 114	112	19	2,128	238,336
115 – 119	117	10	1,170	136,890
120 – 124	122	7	854	104,188
125 – 129	127	2	254	32,258
130 – 134	132	1	132	17,424
Totals		50	5,705	652,945

$$\bar{x} = \frac{5,705}{50} = 114.1° \text{ F}, \ S_{xx} = 652,945 - \frac{(5,705)^2}{50} = 2,004.5, \text{ and } s = \sqrt{\frac{2,004.5}{49}} \approx 6.4° \text{ F}.$$

3.95   Practice exercise.

3.97

Sq. Miles	x	f	xf	Cumulative f
100 – 199	149.50	2	299.0	2
200 – 299	249.50	17	4,241.50	19
300 – 399	349.50	17	5,941.50	36
400 – 499	449.50	20	8,990.00	56
500 – 599	549.50	19	10,440.50	75
600 – 699	649.50	10	6,495.00	85
700 – 799	749.50	5	3,747.50	90
800 – 899	849.50	6	5,097.00	96
900 – 999	949.50	4	3,798.00	100
Totals		100	49,050.00	

(a) $\bar{x} = \dfrac{49,050}{100} = 490.50$ sq. miles.

(b)   Location of the median is the $\dfrac{100}{2} = $ 50th value which, counting from the lowest value, is located in the 400 – 499 class, and its value is $399.5 + \dfrac{14}{20} \cdot 100 = 469.5$ sq. miles.

3.99   Practice exercise.

3.101

PE Ratio	x	f	xf	x^2f	Cumulative f
0 – 4	2	3	6	12	3
5 – 9	7	10	70	490	13
10 – 14	12	12	144	1,728	25
15 – 19	17	9	153	2,601	34
20 – 24	22	4	88	1,936	38
25 – 28	27	2	54	1,458	40
Totals		40	515	8,225	

(a) $\bar{x} = \dfrac{515}{40} \approx 12.9$;

(b) $S_{xx} = 8,225 - \dfrac{(515)^2}{40} \approx 1,594.375$, $s = \sqrt{\dfrac{1,594.375}{40-1}} \approx 6.4$

(c) For median, the location is $\dfrac{40}{2} = 20$, which is 7 items inside the 10 – 14 class.

   Thus, $\tilde{x} = 9.5 + \dfrac{7}{12}(5) \approx 12.4$;

(d) For Q_1 the location is $\left(\dfrac{1}{4}\right)(40) = 10$, which is 7 values inside the 5 – 9 class.

   Thus, $Q_1 = 4.5 + \dfrac{7}{10}(5) = 8.0$;

(e) For Q_3 the location is $\left(\dfrac{3}{4}\right)(40) = 30$, which is 5 values inside the 15 – 19 class.

   Thus $Q_3 = 14.5 + \dfrac{5}{9}(5) \approx 17.3$;

(f) For P_{80}, the location is $(0.80)(40) = 32$, which is 7 values inside the 15 – 19 class.

   Thus $P_{80} = 14.5 + \dfrac{7}{9}(5) \approx 18.4$.

3.103 (a) Symmetrical. Median is in center of box. Tails are of identical length.
   (b) Positive skewness. Median is to the left of center of box. Long whisker on right side.
   (c) Negative skewness. Median is to the right of center of box. Long whisker on left side.

3.105 $P_{20} = 59.5 + \dfrac{12}{26} \cdot 10 \approx 64.12\%$.

   $P_{80} = 79.5 + \dfrac{30}{45} \cdot 10 \approx 86.17\%$.

3.107 Practice exercise.

3.109 $s = \sqrt{0.23} = 0.48$, and $SK = \dfrac{3(2.25 - 1.96)}{0.48} \approx 1.81$.

3.111 The distribution is U-shaped because the number of H's is apt to stay ahead of the number of T's once it gets ahead, and vice versa.

3.113 (a) $x_1 + x_2 + x_3 + x_4 + x_5 + x_6$;

     (b) $y_1 + y_2 + y_3 + y_4 + y_5$;

     (c) $x_1 y_1 + x_2 y_2 + x_3 y_3$;

     (d) $x_1 f_1 + x_2 f_2 + x_3 f_3 + x_4 f_4 + x_5 f_5 + x_6 f_6 + x_7 f_7 + x_8 f_8$;

     (e) $x_3^2 + x_4^2 + x_5^2 + x_6^2 + x_7^2$;

     (f) $(x_1 + y_1) + (x_2 + y_2) + (x_3 + y_3) + (x_4 + y_4)$.

3.115 Practice exercise.

3.117 (a) $2 + 3 + 4 + 5 + 6 + 7 = 27$;

     (b) $3 + 12 + 10 + 6 + 3 + 1 = 35$;

     (c) $2 \cdot 3 + 3 \cdot 12 + 4 \cdot 10 + 5 \cdot 6 + 6 \cdot 3 + 7 \cdot 1 = 137$;

     (d) $4 \cdot 3 + 9 \cdot 12 + 16 \cdot 10 + 25 \cdot 6 + 36 \cdot 3 + 49 \cdot 1 = 587$.

3.119 (a) $2 + 1 + 2 + 3 = 8$;

           $3 + 2 - 2 - 4 = -1$;

           $-1 + 2 + 2 - 3 = 0$.

     (b) $2 + 3 - 1 = 4$;

           $1 + 2 + 2 = 5$;

           $2 - 2 + 2 = 2$;

           $3 - 4 - 3 = -4$.

     (c) $2 + 3 - 1 + 1 + 2 + 2 + 2 - 2 + 2 + 3 - 4 - 3 = 7$

3.121 No.

R.1   (a) $\bar{x} = \dfrac{2,952}{16} = 184.5$ home runs.

       (b) $\dfrac{n+1}{2} = \dfrac{16+1}{2} = 8.5$, and the mean of the 8th and 9th values in the arranged data is

           $\dfrac{194+176}{2} = 185$. Therefore $\tilde{x} = 185$.

R.3   Statements (a) and (d) are descriptive, while statements (b) and (c) are generalizations.

R.5   (a)   The data would constitute a <u>population</u> if the C.P.A. wanted to determine the average Federal Income Tax withheld from all the paychecks of residents of N.Y.C. in the year 2003 only.

       (b)   The data would constitute a <u>sample</u> if the C.P.A. wants to predict the average Federal Income Tax which will be withheld in subsequent years from the residents of New York City or other major cities.

R.7   The mean is not a good average in this case. The calculated value of 14.2 is greatly enlarged by one large value from California and is not at all like the values of the other states.

R.9   Aside from the two families with incomes of $180,000 and $175,000, the mean of $72,900 is higher than the incomes of all other families. The mean is not a central value in this case.

R.11

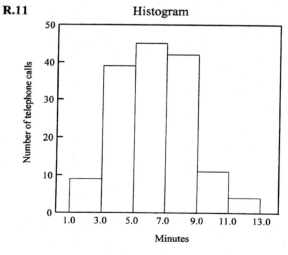

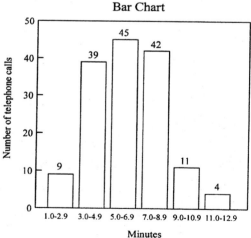

R.13 (a) Find $\sum x = 0.510$ and $\sum x^2 = 0.032646$. For part (a) find $\bar{x} = \dfrac{0.510}{8} = 0.06375$ and

$$S_{xx} = 0.032646 - \frac{0.510^2}{8} = 0.0001335.$$

Then $s = \sqrt{\dfrac{0.0001335}{7}} \approx \sqrt{0.00001907} \approx 0.0044$ p.p.m.

(b) The median is 0.0625 p.p.m., and the range is 0.012 p.p.m.

R.15
```
3 | 2 5 5 5 7
4 | 1 8 4 7 3 9 1 1 1 1 4 9
5 | 7 2 4 9 4 9 2 7
6 | 2 8 4 8
7 | 3
```

R.17 (a) Few persons will give honest answers about their personal habits.
(b) Because of the holiday season, December spending is not typical of spending patterns throughout the year.

R.19 The average arrival time is 8:34 a.m. The interval in question is $8:34 \pm 0:09$, and 9 minutes represent 4 standard deviations. By Chebyshev's theorem, the probability is at least

$$1 - \frac{1}{4^2} = \frac{15}{16}.$$

R.21 (a) Since $\dfrac{38+1}{2} = 19.5$, the median is the mean of the 19th ad 20th values in the group which has been arranged according to size.

$$\tilde{x} = \frac{1{,}595 + 1{,}550}{2} = 1{,}572.5 \text{ feet.}$$

(b) Since the median is the 19.5 value, the location of Q_1 is $\dfrac{19+1}{2} = 10$, and its value is 1,150. The location of Q_3 is 10 from the other end, and its value is 2,150. Since the values are arranged from the largest to the smallest, the designations Q_1 and Q_3 can be reversed.

R.23 (a) 50, 100, 150, 200, 250, 300, and 350;
(b) 99, 149, 199, 249, 299, 349, and 399;
(c) 49.5, 99.5, 149.5, 199.5, 249.5, 299.5, 349.5, and 399.5;
(d) 74.5, 124.5, 174.5, 224.5, 274.5, 324.5, and 374.5;
(e) 50.

R.25 Since $n = 10$, $\sum x = 342$, and $\sum x^2 = 12{,}212$, we get $S_{xx} = 12{,}212 - \dfrac{(342)^2}{10} = 515.6$, and

$$s = \sqrt{\frac{515.6}{9}} \approx 7.6 \text{ work stoppages.}$$

R.27 (a) It is possible to find both the mean and the median.
(b) It is possible to find the median which falls into the $100 - 109$ class, but not the mean.
(c) It is impossible to find either the median or the mean.

R.29   (a)   This is the histogram:

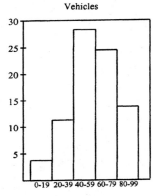

Vehicles

(b)   This is the cumulative "less than" distribution:

Number of Vehicles	Cumulative Frequency
Less than 20	4
Less than 40	15
Less than 60	43
Less than 80	67
Less than 100	80

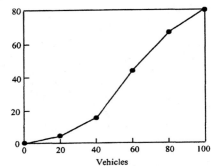

R.31   Here is a possible drawing for the box-and-whisker plot.

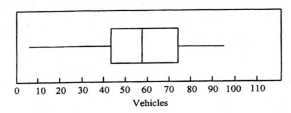

R.33   (a) $\bar{x} = \dfrac{142.5 + 182.1 + 268.2 + 139.6 + 214.4}{5} \approx \$189.4$ thousands.

Arranging the data according to size, we get 139.6, 142.5, 182.1, 214.4, 268.2. The

location of the median is $\dfrac{n+1}{2} = \dfrac{5+1}{2} = 3$, and the third value of the arranged data is $\tilde{x}$

$= 182.1$.

(b) $\bar{x} \approx \$308.2$ thousands, and $\tilde{x}$ is unchanged.

(c)   The printing error seriously affected the mean, but the median was unaffected.

R.35   (a)

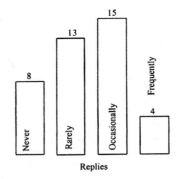

(b)   The modal reply is "occasionally."

<u>With rearranged leaves</u>

R.37	21	6	
	22	4 8 9 1 8 3 2	
	23	5 5 0 6 7 3 1 0 0	
	24	0 4 9 0 5	
	25	3 6	
	26	9	

21	6
22	1 2 3 4 8 8 9
23	0 0 0 1 3 5 5 6 7
24	0 0 4 5 9
25	3 6
26	9

R.39   Location of the median is $\dfrac{n+1}{2} = \dfrac{50+1}{2} = 25.5$ values and $\tilde{x} = \dfrac{17+18}{2} = 17.5$ millions of

dollars. Location of Q_1 is $\dfrac{25+1}{2} = 13$, and counting 13 values from the low end of the

distribution we find $Q_1 = 12$ millions of dollars. Then counting 13 values from the high end

of the distribution we find $Q_3 = 26$ millions of dollars.

Chapter 4

Possibilities and Probabilities

4.1   Practice Exercise.

4.3   The tree diagram for this exercise is

In two cases he will be exactly $1 ahead.

4.5   a. $5 \cdot 5 \cdot 5 \cdot 5 \cdot 5 \cdot 5 = 5^6 = 15,625$.

     b.   1.

     c. $4 \cdot 4 \cdot 4 \cdot 4 \cdot 4 \cdot 4 = 4^6 = 4,096$.

4.7   Practice exercise.

4.9 $5 \cdot 4 \cdot 3 = 60$ ways.

4.11 $20 \cdot 25 = 500$ different ways.

4.13 $5 \cdot 4 \cdot 3 \cdot 2 = 120$ ways.

4.15 $26 \cdot 26 \cdot 10 \cdot 10 \cdot 10 \cdot 10 \cdot 10 = 67,600,000$ identifiable charge cards.

4.17 $3 \cdot 20 = 60$ ways.

4.19   Here is one possible version of this diagram. The symbols are used as follows:
SK = spackle the kitchen
PK = paint the kitchen
SD = spackle the dining room
PD = paint the dining room

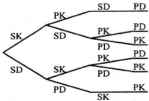

There are six different ways in which these tasks can be done.

4.21   Practice exercise.

4.23   (a)   Since $\frac{2}{6} + \frac{1}{6} = \frac{1}{2}$, the answer is false.

           (b)   Since $\frac{15!}{12!} = 15 \cdot 14 \cdot 13$, the answer is true.

           (c)   Since $\frac{6 \cdot 5 \cdot 4 \cdot 3 \cdot 2 \cdot 1}{3 \cdot 2 \cdot 1 \cdot 1 \cdot 2 \cdot 1} = 60$, the answer is false.

4.25   Practice exercise.

4.27   a. $_{10}P_3 = \dfrac{10!}{(10-3)!} = \dfrac{10!}{7!} = 10 \cdot 9 \cdot 8 = 720$ permutations.

          b. $_3P_2 = \dfrac{3!}{(3-2)!} = \dfrac{3!}{1!} = 6$ permutations.

          c.

1. Miziuko	Citigroup	Allianz
2. Miziuko	Allianz	Citigroup
3. Citigroup	Miziuko	Allianz
4. Citigroup	Allianz	Miziuko
5. Allianz	Miziuko	Citigroup
6. Allianz	Citigroup	Miziuko

4.29   For (a), (b), and (c), the solutions are all $6 \cdot 5 \cdot 4 \cdot 3 \cdot 2 \cdot 1 = 720$ ways.

4.31   (a)   To find the number of ways in which n objects can be arranged in a circle, we arbitrarily fix the position of one of the objects and calculate the number of ways in which the remaining positions can be filled. Thus, the number of circular permutations is $(n-1)!$.

          (b) $5! = 120$;

          (c) $3! = 6$.

4.33   (a) $\dfrac{7!}{3!} = 840$.

          (b) $\dfrac{8!}{3! \cdot 2!} = 3{,}360$.   It is necessary to deal with the three o's and the two l's.

          (c) $\dfrac{9!}{2! \cdot 2!} = 90{,}720$.

          (d) $\dfrac{11!}{4! \cdot 4! \cdot 2!} = 34{,}650$.

4.35   Practice exercise.

4.37 (a) $\dfrac{14 \cdot 13 \cdot 12}{3 \cdot 2 \cdot 1} = 364$ (b) Identical to (a)

(c) $\dfrac{15 \cdot 14 \cdot 13 \cdot 12 \cdot 11}{5 \cdot 4 \cdot 3 \cdot 2 \cdot 1} = 3{,}003$ (d) Identical to (c)

4.39 (a) $\dbinom{25}{5}$ (c) $\dbinom{10}{3}$

(b) $\dbinom{14}{4}$ (d) $\dbinom{18}{6}$

4.41 (a) $\dbinom{8}{3} \cdot \dbinom{4}{0} = 56 \cdot 1 = 56;$ (c) $\dbinom{8}{1} \cdot \dbinom{4}{2} = 8 \cdot 6 = 48;$

(b) $\dbinom{8}{2} \cdot \dbinom{4}{1} = 28 \cdot 4 = 112;$ (d) $\dbinom{8}{0} \cdot \dbinom{4}{3} = 1 \cdot 4 = 4.$

4.43 Reading the answers directly from Table X, Binomial Coefficients, we get carpenters, 10 ways; plumbers, 6 ways; and electricians, 15 ways. The overall result is obtained by multiplying the 3 values, obtaining $10 \cdot 6 \cdot 15 = 900$ ways.

4.45 Reading directly from the Table of Binomial Coefficients we get, for parts (a) and (b):

(a) 6 ways;

(b) 10 ways;

(c) $6 \times 10 = 60$

4.47 The next three rows of Pascal's triangle are
    1 6 15 20 15 6 1
  1 7 21 35 35 21 7 1
 1 8 28 56 70 56 28 8 1

4.49 Practice exercise.

4.51 (a) $\dfrac{2}{52} = \dfrac{1}{26};$ (c) $\dfrac{16}{52} = \dfrac{4}{13};$ (e) $\dfrac{39}{52} = \dfrac{3}{4};$

(b) $\dfrac{26}{52} = \dfrac{1}{2};$ (d) $\dfrac{13}{52} = \dfrac{1}{4};$ (f) $\dfrac{48}{52} = \dfrac{12}{13}.$

4.53 The probabilities are $\dfrac{1}{8}, \dfrac{3}{8}, \dfrac{3}{8},$ and $\dfrac{1}{8}.$

4.55 (a) $\dfrac{1}{6}$;

(b) The even numbers are 2, 4, and 6, so the probability is $\dfrac{3}{6}$ or $\dfrac{1}{2}$.

4.57 The 3 good flashlights can be selected in $\dbinom{15}{3}$ ways and the defective flashlights in $\dbinom{4}{2}$ ways, and by multiplication of choices we have $\dbinom{15}{3} \cdot \dbinom{4}{2} = 455 \cdot 6 = 2{,}730$ ways. (The binomial coefficients can be found in Table X.)

4.59 (a) $\dfrac{\dbinom{10}{3}}{\dbinom{30}{3}} = \dfrac{120}{4{,}060} = \dfrac{6}{203}$ (b) $\dfrac{\dbinom{20}{2}\dbinom{10}{1}}{\dbinom{30}{3}} = \dfrac{190\cdot10}{4{,}060} = \dfrac{95}{203}$.

4.61 Practice exercise.

4.63 $\dfrac{39{,}865{,}000}{72{,}455{,}000} \approx 0.55$

4.65 $\dfrac{1{,}035{,}000}{5{,}176{,}000} = 0.20$

4.69 Practice exercise.

4.71 $0 \cdot \dfrac{299}{300} + 30 \cdot \dfrac{1}{300} = 0 + 0.10 = \$0.10$, or 10¢.

4.73 Begin by noting that the probability of getting a red number is $\dfrac{18}{38} = \dfrac{9}{19}$, the probability of getting a black number is $\dfrac{18}{38} = \dfrac{9}{19}$, and the probability of getting a green number is $\dfrac{2}{38} = \dfrac{1}{19}$.

(a) $(+\$1)\dfrac{18}{38} + (-\$1)\dfrac{20}{38} \approx -\$0.0526$;   (b) $(+\$2)\dfrac{12}{38} + (-\$1)\dfrac{26}{38} \approx -\$0.0526$;

(c) $(+\$35)\dfrac{1}{38} + (-\$1)\dfrac{37}{38} \approx -\$0.0526$.

4.75 $0.10(100{,}000) + 0.25(90{,}000) + 0.35(80{,}000) + 0.30(70{,}000)$
$= 10{,}000 + 22{,}500 + 28{,}000 + 21{,}000 = \$81{,}500$.

4.77 A child in this age group can expect to visit a dentist
$0(0.15) + 1(0.28) + 2(0.27) + 3(0.17) + 4(0.08) + 5(0.03) + 6(0.02) = 1.92$ times in any given year.

4.79 Practice exercise.

4.81   If the probability that the race will have to be canceled because of rain is p, we have

$15,000p = 2,400$, so that $p = \dfrac{2,400}{15,000} = 0.16$.

4.83 $0.25(0) + 0.35(1) + 0.24(2) + 0.12(3) + 0.04(4)$
$= 0 + 0.35 + 0.48 + 0.36 + 0.16 = 1.35$ lawsuit filings.

4.85   Practice exercise.

4.87   The expected profit is $4,500,000 \cdot \dfrac{1}{2} - 2,700,000 \cdot \dfrac{1}{2} = \$900,000$ if they continue the operation

and $-1,800,000 \cdot \dfrac{1}{2} + 450,000 \cdot \dfrac{1}{2} = -\$675,000$ if they do not continue the operation. Thus, continuing the operation will maximize the company's expected profit.

4.89   The expected profit is $451,000 \cdot \dfrac{1}{6} - 110,000 \cdot \dfrac{5}{6} = -\$16,500$ if the new factory is built, and

$220,000 \cdot \dfrac{1}{6} + 22,000 \cdot \dfrac{5}{6} = \$55,000$ if the new factory is not built. Thus, not building the new factory will maximize the expected profit.

4.91   The profit table (in thousands) is the following:

		Number of Tenants				
		0	1	2	3	4+
	0	0	0	0	0	0
Number of stores	1	−12	18	18	18	18
fitted with	2	−24	6	36	36	36
indoor furnishings	3	−36	−6	24	54	54
	4	−48	−18	12	42	72

Using the probabilities provided by the expert, we compute the expected profit for each number of stores fitted with indoor furnishings.

	0	$0 \cdot 0.1 +$	$0 \cdot 0.4 +$	$0 \cdot 0.3 +$	$0 \cdot 0.1 +$	$0 \cdot 0.1 =$	0.0
	1	$-12 \cdot 0.1 +$	$18 \cdot 0.4 +$	$18 \cdot 0.3 +$	$18 \cdot 0.1 +$	$18 \cdot 0.1 =$	15.0
Number of	2	$-24 \cdot 0.1 +$	$\cdots \cdot 0.4 +$	$36 \cdot 0.3 +$	$36 \cdot 0.1 +$	$36 \cdot 0.1 =$	18.0
stores	3	$-36 \cdot 0.1 -$	$6 \cdot 0.4 +$	$24 \cdot 0.3 +$	$54 \cdot 0.1 +$	$54 \cdot 0.1 =$	12.0
fitted with	4	$-48 \cdot 0.1 -$	$18 \cdot 0.4 +$	$12 \cdot 0.3 +$	$42 \cdot 0.1 +$	$72 \cdot 0.1 =$	3.0
indoor							
furnishings							

The best strategy calls for furnishing two stores, and the expected profit is $18,000.

Chapter 5

Some Rules of Probability

5.1 Practice exercise.

5.3 $D' = \{1,2,6\}$ which is the event we roll a 1, 2, or 6.
$D \cup E = \{2,3,4,5,6\}$ which is the event we do not roll a 1.
$D \cap E = \{4\}$ which is the event we roll a 4.

5.5 (a) $M \cup N = \{Q\clubsuit, K\clubsuit, Q\diamondsuit, K\diamondsuit, Q\heartsuit, K\heartsuit, 10\spadesuit, J\spadesuit, Q\spadesuit, K\spadesuit\}$ is the event that we draw a queen, a king, or the 10 or jack of spades.
(b) $M \cap N = M \cup N = \{Q\spadesuit, K\spadesuit\}$ is the event that we draw the queen or king of spades.
(c) $M' = (A\clubsuit, 2\clubsuit,..., J\clubsuit, A\diamondsuit, 2\diamondsuit,..., J\diamondsuit, A\heartsuit, 2\heartsuit,..., J\heartsuit, A\spadesuit, 2\spadesuit,..., J\spadesuit)$ is the event that we do not draw a queen or a king.
(d) $N' = \{A\clubsuit,..., K\clubsuit, A\diamondsuit,..., K\diamondsuit, A\heartsuit,..., K\heartsuit, A\spadesuit,..., 9\spadesuit\}$ is the event that we do not draw the 10, jack, queen, or king of spades.
(e) $M' \cup N' = \{A\clubsuit,..., K\clubsuit, A\diamondsuit,..., K\diamondsuit, A\heartsuit,..., K\heartsuit, A\spadesuit,..., J\spadesuit\}$ is the event that we do not draw the queen or king of spades.
(f) $M' \cap N' = \{A\clubsuit,..., J\clubsuit, A\diamondsuit,..., J\diamondsuit, A\heartsuit,..., J\heartsuit, A\spadesuit,..., 9\spadesuit\}$ is the event that we do not draw a queen, a king, the 10 of spades or the jack of spades.

5.7 Since $A \cap B = \{3,4\}$, A and B are not mutually exclusive.
Since $A \cap C = \varnothing$, A and C are mutually exclusive.
Since $B \cap C = \{5,6,7\}$, B and C are not mutually exclusive.

5.9 (a) Mutually exclusive;
(b) Not mutually exclusive
(c) Not mutually exclusive;
(d) Mutually exclusive;

5.11 (a) Not mutually exclusive;
(b) mutually exclusive;
(c) not mutually exclusive;
(d) not mutually exclusive;
(e) not mutually exclusive;
(f) mutually exclusive.

5.13 (a) Regions 1 and 2 together represent the event that a person vacationing in Southern California visits Disneyland.
(b) Regions 2 and 3 together represent the event that a person vacationing in Southern California visits Disneyland or Universal Studios, but not both.
(c) Regions 2 and 4 together represent the event that a person vacationing in Southern California does not visit Universal Studios.

5.15 (a) Regions 1 and 3 together represent the event that the murder suspect is guilty.
(b) Regions 1 and 4 together represent the event that either the murder suspect is guilty and allowed out on bail, or is not guilty and not allowed out on bail.

(c) Regions 3 and 4 together represent the event that the murder suspect is not allowed out on bail.

5.17 (a) The laboratory facilities are adequate, the financial support is adequate, and it is staffed by competent personnel.

(b) The financial support is adequate, it is staffed by competent personnel, but the laboratory facilities are not adequate.

(c) It is staffed by competent personnel, but the laboratory facilities and financial support are not adequate.

(d) The laboratory facilities are not adequate, the financial support is not adequate, and it is not staffed by competent personnel.

(e) The laboratory facilities are adequate and it is staffed by competent personnel.

(f) The financial support is adequate, it is staffed by competent personnel, but the laboratory facilities are not adequate.

(g) The personnel are competent.

(h) The personnel are not competent.

5.19 Practice exercise.

5.21 (a) $0.10 + 0.15 + 0.20 + 0.25 = 0.70$;

(b) $1.00 - (0.10 + 0.15) = 0.75$;

(c) $0.20 + 0.25 = 0.45$

5.23 (a) The probability cannot be greater than 1, and it is given as 1.09.

(b) The sum of the two probabilities cannot be greater than 1, and it is 1.10.

(c) The sum of the two probabilities is $0.30 + 0.40 = 0.70$, and not 0.90.

(d) The sum of the two probabilities should be 1, but it is given as $0.55 + 0.35 = 0.90$.

5.25 (a) $1 - 0.45 = 0.55$; (d) 0;

(b) $1 - 0.30 = 0.70$; (e) $P(Q' \cap R) = P(R) = 0.30$;

(c) $0.45 + 0.30 = 0.75$; (f) $P(Q \cap R') = P(Q) = 0.45$.

5.27 (a) $1 - (0.55 + 0.10) = 0.35$

(b) $0.35 + 0.10 = 0.45$

(c) $0.55 + 0.10 = 0.65$

5.29 1. Since $0 \leq s \leq n$, division by n yields $0 \leq \frac{s}{n} \leq 1$.

2. If an event is certain to occur, $s = n$, so that its probability is $\frac{s}{s} = \frac{n}{n} = 1$; if an event is certain not to occur, then $s = 0$ and the probability is 0.

3. The respective probabilities are $\frac{s_1}{n}, \frac{s_2}{n}$, and $\frac{s_1 + s_2}{n}$, and the sum of the first two is equal to the third.

4. $\frac{s}{n} + \frac{n-s}{n} = \frac{n}{n} = 1$.

5.31 Practice exercise.

5.33   (a)   The odds are $\dfrac{6}{10}$ to $\left(1 - \dfrac{6}{10}\right)$ = 6 to 4 or 3 to 2;

      (b)   The odds are $\dfrac{8}{24}$ to $\left(1 - \dfrac{8}{24}\right)$ = 8 to 16 or 1 to 2, which is 2 to 1 against the occurrence;

      (c)   The odds are $\dfrac{6}{16}$ to $\left(1 - \dfrac{6}{16}\right)$ = 6 to 10 or 3 to 5, which is 5 to 3 against the occurrence;

      (d)   The odds are $\dfrac{13}{52}$ to $\left(1 - \dfrac{13}{52}\right)$ or 13 to 39 or 1 to 3, which is 3 to 1 against the occurrence.

5.35   The probabilities that he will run for the House of Representatives, the Senate, or either one, are $\dfrac{1}{1+5} = \dfrac{1}{6}, \dfrac{1}{1+4} = \dfrac{1}{5}$, and $\dfrac{1}{1+2} = \dfrac{1}{3}$, and since $\dfrac{1}{6} + \dfrac{1}{5} = \dfrac{11}{30}$, which is more than equal to $\dfrac{1}{3}$, they are not consistent.

5.37   The demonstration proceeds as follows:

$$a(1 - p) = bp$$
$$a - ap = bp$$
$$a = ap + bp$$
$$a = p(a + b)$$
$$p = \dfrac{a}{a + b}.$$

5.39   (a)   The even money bet is attractive.
      (b)   Attractive.
      (c)   Unattractive.

5.41 $1.00 - (0.08 + 0.23 + 0.45) = 0.24$

5.43   (a) $0.19 + 0.18 + 0.14 + 0.09 = 0.60$;
      (b) $0.02 + 0.04 + 0.07 + 0.12 = 0.25$;
      (c) $0.07 + 0.12 + 0.15 = 0.34$.

5.45   Practice exercise.

5.47   (a) $0.17 + 0.17 = 0.34$;
      (b) $0.22 + 0.22 = 0.44$;
      (c) $0.06 + 0.16 = 0.22$.

5.49   Practice exercise.

5.51 $0.06 + 0.08 - 0.03 = 0.11$

5.53　Filling the numbers associated with the various regions, we get

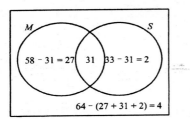

and it can be seen that the probability is $\dfrac{4}{64} = \dfrac{1}{16}$.

5.55　Practice exercise.

5.57　(a)　$P(X|Y) = P(X) = 0.25$;
　　　　(b)　$P(X \cap Y) = P(X) \cdot P(Y) = 0.25 \cdot 0.50 = 0.125$;
　　　　(c)　$P(X \cup Y) = P(X) + P(Y) = 0.25 + 0.50 = 0.75$;
　　　　(d)　$P(X' \cap Y') = 1 - 0.625 = 0.375$.

5.59　$P(A) \cdot P(B) = (0.70)(0.40) = 0.28 \neq 0.25$, so that A and B are not independent.

5.61　Practice exercise.

5.63　(a)　$\dfrac{240}{340} = \dfrac{12}{17} \approx 0.706$;　　　　(d)　$\dfrac{57}{340} \approx 0.168$;

　　　　(b)　$\dfrac{115}{340} = \dfrac{23}{68} \approx 0.338$;　　　　(e)　$\dfrac{168}{225} = \dfrac{56}{75} \approx 0.747$;

　　　　(c)　$\dfrac{72}{340} = \dfrac{18}{85} \approx 0.212$;　　　　(f)　$\dfrac{57}{100} = 0.570$.

5.65　(a)　$\dfrac{12 + 18 - 5}{60} = \dfrac{5}{12}$;　　　　(d)　$\dfrac{1 + 3}{11} = \dfrac{4}{11}$;

　　　　(b)　$\dfrac{3}{23}$;　　　　(e)　$\dfrac{10 + 9}{23 + 19} = \dfrac{19}{42}$;

　　　　(c)　$\dfrac{9}{60} = \dfrac{3}{20}$;　　　　(f)　$\dfrac{3 + 5}{11 + 12} = \dfrac{8}{23}$.

5.67　(a)　The probability that a very rare sword is in good condition.
　　　　(b)　The probability that an antique sword which is not a reproduction is in good condition.
　　　　(c)　The probability that an antique sword which is not very rare is a reproduction.
　　　　(d)　The probability that a very rare antique sword is in good condition, and is a reproduction.
　　　　(e)　The probability that an antique sword which is not in good condition is very rare.
　　　　(f)　The probability that an antique sword which is very rare and a reproduction is in good condition.

5.69 (a) $\dfrac{1}{2} \cdot \dfrac{1}{2} \cdot \dfrac{1}{2} \cdot \dfrac{1}{2} \cdot \dfrac{1}{2} = \dfrac{1}{32}$

   (b) $\dfrac{1}{13} \cdot \dfrac{1}{13} \cdot \dfrac{1}{13} = \dfrac{1}{2,197}$

   (c) $\dfrac{13}{52} \cdot \dfrac{12}{51} \cdot \dfrac{11}{50} = \dfrac{1,716}{132,600} \approx 0.013$

   (d) $(0.80)(0.20) = 0.16$

5.71 (a) In the first game, Abel's proportion of completions was $\dfrac{11}{20} = 0.55$, while Baker's was

   $\dfrac{3}{5} = 0.60$. Baker did better.

   (b) In the second game, Abel's proportion of completions was $\dfrac{1}{5} = 0.20$, while Baker's was

   $\dfrac{6}{16} = 0.375$. Baker did better again.

   (c) For the season-to-date, Abel threw 25 passes, of which 12 were completions. His
   proportion of completions was 0.48. Baker threw 21 passes, with 9 completions, so his
   proportion of completions was approximately 0.4286. Thus, even though Baker did
   better in each game, his overall performance was worse!

5.73 $P(L|A) = \dfrac{0.63}{0.75} = 0.84 = P(L)$.

5.75 (a) $\dfrac{1}{6} \cdot \dfrac{1}{6} = \dfrac{1}{36}$;

   (b) $\dfrac{1}{6} \cdot \dfrac{1}{6} = \dfrac{1}{36}$
   The results of rolling a single die 2 times, or 2 dice 1 time.

5.77 (a) $\dfrac{16}{30} \cdot \dfrac{16}{30} = \dfrac{64}{225}$;   (b) $\dfrac{16}{30} \cdot \dfrac{15}{29} = \dfrac{8}{29}$.

5.79 (a) $0.35 + 0.25 - 0.15 = 0.45$

5.81 (a) $\dfrac{10}{24} \cdot \dfrac{9}{23} \cdot \dfrac{8}{22} \approx (0.42) \cdot (0.39) \cdot (0.36) \approx 0.06$;

   (b) $\dfrac{14}{24} \cdot \dfrac{13}{23} \cdot \dfrac{12}{22} \approx 0.18$

5.83 (a) $(0.4)(0.7)(0.7) = 0.196$;   (b) $(0.6)(0.2)(0.7) = 0.084$.

5.85 Practice exercise.

5.87   Let A, B, and C correspond to Arthur, Beatrice, and Carla in the following diagram.

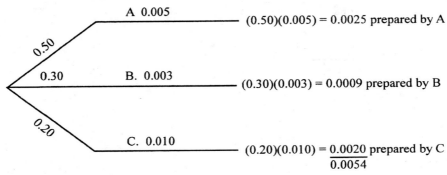

A 0.005 $(0.50)(0.005) = 0.0025$ prepared by A

0.50

0.30   B. 0.003 $(0.30)(0.003) = 0.0009$ prepared by B

0.20

C. 0.010 $(0.20)(0.010) = \underline{0.0020}$ prepared by C
                                                 0.0054

and

(a) $\dfrac{0.0025}{0.0054} \approx 0.46$

(b) $\dfrac{0.0009}{0.0054} \approx 0.17$

(c) $\dfrac{0.0020}{0.0054} \approx 0.37$

5.89 $P(B|A) = \dfrac{(0.75)(0.85)}{(0.75)(0.85) + (0.25)(0.60)} \approx 0.810.$

5.91   If T and Y denote the events that one of the cars needs a tune-up and that it comes from rental agency Y, then $P(Y|T) = \dfrac{(0.25)(0.06)}{(0.25)(0.08) + (0.25)(0.06) + (0.50)(0.15)} = \dfrac{0.015}{0.110} \approx 0.136.$

R.41 $\begin{pmatrix} 5 \\ 3 \end{pmatrix} \cdot \begin{pmatrix} 8 \\ 3 \end{pmatrix} \cdot \begin{pmatrix} 7 \\ 3 \end{pmatrix} = 10 \cdot 56 \cdot 35 = 19{,}600$ ways.

R.43 The 15 points of the sample space are shown in the following diagram:

R.45 The expected number is $0(0.10) + 1(0.43) + 2(0.29) + 3(0.18) = 1.55$ alarms

R.47 (a) $\{A, D\}$;
　　(b) $\{C, E\}$;
　　(c) $\{B\}$.

R.49 Altogether the 1,000 raffle tickets will pay $500 + $100 + $50 or, on the average, $0.65 per ticket. Since the ticket costs $1.00, it is not a wise purchase.

R.51 (a) $\dfrac{3}{10} \cdot \dfrac{2}{9} = \dfrac{1}{15}$; 　　　　(b) $\dfrac{7}{10} \cdot \dfrac{6}{9} = \dfrac{7}{15}$

R.53 Create a display giving the name of the check received by each person.

Received by A	J	P	W	Match		Received by A	J	P	W	Match
A	J	P	W	4		P	A	J	W	1
A	J	W	P	2		P	A	W	J	0
A	P	J	W	2		P	J	A	W	2
A	P	W	J	1		P	J	W	A	1
A	W	J	P	1		P	W	A	J	0
A	W	P	J	2		P	W	J	A	0
J	A	P	W	2		W	A	J	P	0
J	A	W	P	0		W	A	P	J	1
J	P	A	W	1		W	J	A	P	1
J	P	W	A	0		W	J	P	A	2
J	W	A	P	0		W	P	A	J	0
J	W	P	A	1		W	P	J	A	0

Here, the column "Match" indicates the number of people getting their own check.

(a) There are $4! = 24$ ways to distribute the checks to the four Davises.
(b) There are 9 ways in which there are zero matches.
(c) There are 8 ways in which there is one match.
(d) There are 6 ways in which there are two matches.
(e) There are 0 ways in which there are three matches.
(f) There is 1 way in which there are four matches.

R.55 (a) $4 \cdot 3 \cdot 4 = 48$;
(b) $3 \cdot 4 = 12$;
(c) $4 \cdot 3 \cdot 2 = 24$.

R.57 (a) $1 - 0.40 = 0.60$;
(b) $0.40 + 0.25 = 0.65$;
(c) $1 - 0.65 = 0.35$.

R.59 $7 \cdot 6 \cdot 5 = 210$ ways.

R.61 The probability that on any one day the commuter will make the trip within 30 minutes is
$$\frac{216}{324} = \frac{2}{3} \text{ or } 0.67.$$

R.63 Observe that $P(A|R') = \frac{160}{310}, P(A \cap R') = \frac{160}{380}, P(R') = \frac{310}{380}$, and that $\dfrac{\frac{160}{380}}{\frac{310}{380}} = \dfrac{160}{310}$.

R.65 (a) $\dfrac{5}{5+2} = \dfrac{5}{7}$
(b) 2 to 5
(c) $\dfrac{2}{2+5} = \dfrac{2}{7}$
(d) $\dfrac{2}{7} + \dfrac{5}{7} = 1$

R.67 Region 1 represents the event that the school's football team is rated among the top twenty by both AP and UPI. Region 2 represents the event that the school's football team is rated among the top twenty by AP but not by UPI. Region 3 represents the event that the school's football team is rated among the top twenty by UPI but not by AP. Region 4 represents the event that the school's football team is rated among the top twenty by neither AP nor UPI.

R.69 (a) $\dfrac{13}{52} \dfrac{13}{52} \dfrac{13}{52} \dfrac{13}{52} = \dfrac{1}{256}$ or .0039;
(b) $\dfrac{13}{52} \dfrac{12}{51} \dfrac{11}{50} \dfrac{10}{49} = \dfrac{17,160}{6,497,400}$ or 0.0026.

R.71   (a) $\binom{6}{4} = 15$ ways;

       (b) $\binom{6}{2} \cdot \binom{6}{2} = 225$ ways;

       (c) $\binom{12}{4} = 495$ ways.

R.73   (a)   Region 5.
       (b)   Regions 1 and 2 together.
       (c)   Regions 3, 5, and 6 together.
       (d)   Regions 1, 3, 4, and 6 together.

R.75   (a) $\dfrac{10}{10+1} = \dfrac{10}{11}$ or approximately 0.91;

       (b) $\dfrac{5}{5+3} = \dfrac{5}{8}$ or approximately 0.62;

       (c) $\dfrac{1}{1+1} = \dfrac{1}{2}$ or 0.50

R.77   (a) $0.20 + 0.15 = 0.35$;
       (b) $0.10 + 0.13 + 0.20 = 0.43$;
       (c) $0.30 + 0.12 = 0.42$;
       (d) $0.20 + 0.13 + 0.12 = 0.45$.

R.79 $\dfrac{1}{2} \cdot \dfrac{1}{2} \cdot \dfrac{1}{2} \cdot \dfrac{1}{2} = \dfrac{1}{16}$

R.81   (a) $P(A|B) = \dfrac{0.13}{0.20} = 0.65 = P(A)$;

       (b) $P(A|B') = \dfrac{0.65 - 0.13}{1 - 0.20} = \dfrac{0.52}{0.80} = 0.65 = P(A)$;

       (c) $P(B|A) = \dfrac{0.13}{0.65} = 0.20 = P(B)$;

       (d) $P(B|A') = \dfrac{0.20 - 0.13}{1 - 0.65} = \dfrac{0.07}{0.35} = 0.20 = P(B)$.

Chapter 6

Probability Distributions

6.1   Practice exercise.

6.3   (a)   The values sum to 1.30, so they cannot serve as the values of a probability distribution.
    (b)   The values sum to 0.50, so they cannot serve as the values of a probability distribution.
    (c)   The values all sum to 1.00 and there are no negative values, so they can serve as the values of a probability distribution.
    (d)   Since the values of $f(1)$, $f(2)$, $f(3)$, and $f(4)$ are negative, they cannot serve as the values of a probability distribution.

6.5   Practice exercise.

6.7   Practice exercise.

6.9 $\binom{8}{3}(0.20)^3(1-0.20)^5 = (56)(0.008)(0.3277) \approx 0.147.$

    (The value of $\binom{8}{3}$ is $\dfrac{8 \cdot 7 \cdot 6}{3 \cdot 2 \cdot 1} = 56.$)

6.11 $\binom{7}{7}(0.90)^7(0.10)^0 \approx 0.478$

6.13 $f(0) = \binom{5}{0}(0.30)^0(1-0.30)^5 = (1)(0.70)^5 \approx 0.168$

6.15   (a) $0.092 + 0.157 + \ldots + 0.007 + 0.001 = 0.942$ or by subtraction from $1 - (0.001 + 0.003 + 0.014 + 0.041) = 0.941$. The difference is due to rounding.
    (b) $0.001 + 0.003 = 0.004$;
    (c) $0.041 + 0.092 + 0.157 + 0.207 = 0.497$

6.17   (a) 0.196;
    (b) $0.153 + 0.092 + 0.042 + 0.014 + 0.003 = 0.304$;
    (c) 0.153;
    (d) $0.092 + 0.042 + 0.014 + 0.003 = 0.151.$

6.19   (a) $0.002 + 0.015 + 0.082 = 0.099$;
    (b) $0.311 + 0.187 + 0.047 = 0.545$;
    (c) $0.138 + 0.037 + 0.004 = 0.179.$

6.21   (a) $0.0994 + 0.2017 + 0.2731 + 0.2376 + 0.1206 + 0.0272 = 0.9596$;
    (b) $0.2017 + 0.0994 + 0.0326 + 0.0069 + 0.0008 + 0.0000 = 0.3414.$

6.23   (a) $0.1700 + 0.0916 + 0.0305 + 0.0047 = 0.2968$ (No);

(b) $0.1144 + 0.1643 + 0.1916 + 0.1789 + 0.1304 + 0.0716 + 0.0278 + 0.0068 + 0.0008 = 0.8866$ (No);

(c) $0.0115 + 0.0268 + 0.0536 + 0.0916 + 0.1336 + 0.1651 + 0.1712 + 0.1472 + 0.1030 + 0.0572 + 0.0243 + 0.0074 + 0.0014 + 0.0001 = 0.9940$ (Yes);

(d) $n = 21$.

6.25   Practice exercise.

6.27 $\dfrac{\binom{8}{4}\binom{7}{0}}{\binom{15}{4}} = \dfrac{70 \cdot 1}{1,365} = \dfrac{70}{1,365} \approx 0.051$

6.29 $\dfrac{\binom{10}{2}\binom{8}{2}}{\binom{18}{4}} = \dfrac{(45)(28)}{3,060} \approx 0.412$

6.31   (a) $\dfrac{\binom{3}{0}\binom{7}{3}}{\binom{10}{3}} = \dfrac{(1)(35)}{(120)} \approx 0.292$;

(b) $\dfrac{\binom{3}{1}\binom{7}{2}}{\binom{10}{3}} = \dfrac{(3)(21)}{(120)} \approx 0.525$;

(c) $\dfrac{\binom{3}{2}\binom{7}{1}}{\binom{10}{3}} = \dfrac{(3)(7)}{(120)} \approx 0.175$;

(d) $\dfrac{\binom{3}{3}\binom{7}{0}}{\binom{10}{3}} = \dfrac{(1)(1)}{(120)} \approx 0.008$;

6.33   Practice exercise.

6.35 $\dfrac{\binom{36}{0}\binom{54}{3}}{\binom{90}{3}} = \dfrac{1 \cdot 24,804}{117,480} = 0.211$.   The binomial approximation yields

$\binom{3}{0}(0.40)^0(0.60)^3 = 0.216$, so the error would be $0.216 - 0.211 = 0.005$.

6.37   Using $n = 12$, $p = 0.20$, and $x = 1$, the value in Table I is 0.206.

6.39   (a) Since n is less than 100 the conditions are not satisfied.

   (b) Since $np = 200 \cdot \dfrac{1}{5} = 40$ exceeds 10, the conditions are not satisfied.

   (c) Since $n = 500$, is not less than 100 and $500 \cdot \dfrac{1}{60} \approx 8.33$ is less than 10, the conditions are satisfied.

6.41   Use the Poisson approximation with $n \cdot p = 100(0.019) = 1.9$ and find

$$f(3) = 0.150 \cdot \frac{(1.9)^3}{3!} = 0.150 \cdot \frac{6.8590}{6} \approx 0.171.$$

6.43   Use the Poisson approximation with $n \cdot p = (500)(0.012) = 6$, and find

$$f(2) = e^{-6.0} \cdot \frac{(6.0)^2}{2!} = 0.0025 \cdot \frac{36}{2 \cdot 1} \approx 0.0450,$$

$$f(3) = e^{-6.0} \cdot \frac{(6.0)^3}{3!} = 0.0025 \cdot \frac{216}{3 \cdot 2 \cdot 1} \approx 0.0900.$$

and the sum of the probabilities is $0.0450 + 0.0900 = 0.1350$.

6.45   (a) The binomial approximation to the hypergeometric distribution requires that $n \le (0.05)(a + b)$. The Poisson approximation to the binomial distribution requires that $n \ge 100$ and $np < 10$. Since we use $\dfrac{a}{a+b}$ for p, the required conditions are

   $n \le (0.05)(a + b)$, $n \ge 100$, and $n\dfrac{a}{a+b} < 10$.

   (b) We use $n\dfrac{a}{a+b}$ for np in the Poisson calculation.

6.47 $f(6) = e^{-5} \cdot \dfrac{5^6}{6!} = (0.0067)\dfrac{15{,}625}{6 \cdot 5 \cdot 4 \cdot 3 \cdot 2 \cdot 1} = 0.1454 \approx 0.15$

6.49   Substituting $\lambda = 7$ and $x = 4$ into the formula for 4 reports, we get

$$f(4) = \frac{7^4 \cdot e^{-7}}{4!} = \frac{(2{,}401)(0.0009)}{4 \cdot 3 \cdot 2 \cdot 1} = 0.09 \text{ (the solution to only 4 reports)}$$

   Substituting $\lambda = 7$ and $x = 5$ into the formula for 5 reports, we get

$$f(5) = \frac{7^5 \cdot e^{-7}}{5!} = \frac{(16{,}807)(0.0009)}{5 \cdot 4 \cdot 3 \cdot 2 \cdot 1} \approx 0.128.$$

   Summing $f(4)$ and $f(5)$ we get $0.09 + 0.128 \approx 0.218$.

6.51   Substituting $\lambda = 6$ and $x = 7$ into the formula for the Poisson distribution with parameter λ, we get

$$f(7) = \frac{6^7 \cdot e^{-6}}{7!} = \frac{(279{,}936)(0.0025)}{5{,}040} \approx 0.139$$

6.53   (a) 0.843645;   (b)   0.640552.

6.55   Practice exercise.

6.57   The probability is $\frac{5!}{1!\,2!\,2!}(0.25)^1(0.50)^2(0.25)^2 \approx 0.117$.

6.59   (a) 0.207;

      (b) $\frac{15!}{6!\,9!}(0.40)^6(0.60)^9 \approx 0.207$;

      (c) The multinomial distribution with $k = 2$ is the binomial distribution.

6.61 $\mu = 0(0.05) + 1(0.08) + 2(0.12) + 3(0.25) + 4(0.30) + 5(0.20) = 3.27$.

6.63   (a) $\mu = 0(0.05) + 1(0.20) + 2(0.30) + 3(0.25) + 4(0.15) + 5(0.05) = 2.40$
      (b) First solve for σ^2, getting

$$\sigma^2 = (0 - 2.40)^2(0.05) + (1 - 2.40)^2(0.20) + (2 - 2.40)^2(0.30)$$
$$+ (3 - 2.40)^2(0.25) + (4 - 2.40)^2(0.15) + (5 - 2.40)^2(0.05)$$
$$= 1.5400.$$

      Then, solving for σ, we get $\sigma = \sqrt{1.5400} \approx 1.24$.

6.65   (a) $\sigma^2 = (1-3.5)^2 \cdot \frac{1}{6} + (2-3.5)^2 \cdot \frac{1}{6} + \ldots + (6-3.5)^2 \cdot \frac{1}{6} \approx 2.9167$ and $\sigma \approx \sqrt{2.9167} \approx 1.7078$;

      (b) $\sum x^2 \cdot f(x) = 15.1667 - (3.5)^2 = 2.9167$ and $\sigma^2 = \sqrt{2.9168} \approx 1.7078$.

6.67   Practice exercise.

6.69   (a) $\sum x \cdot f(x) = 2.5$ and $\sum x^2 \cdot f(x) = 7.5$, so that $\sigma^2 = 7.5 - (2.5)^2 = 1.25$;
      (b) $\sigma^2 = 5(0.5)(0.5) = 1.25$.

6.71   (a) $\sigma^2 = (5 - 9)^2(0.001) + (6 - 9)^2(0.011) + (7 - 9)^2(0.057) + (8 - 9)^2(0.194) +$
         $(9 - 9)^2(0.387) + (10 - 9)^2(0.349) = 0.886$, and $= \sqrt{0.886} \approx 0.9413$;
      (b) $\sqrt{10(0.90)(1 - 0.90)} = \sqrt{0.90} \approx 0.9487$.

6.73   (a) $(5)(0.50) = 2.5$ heads.
      (b) $(10)(0.05) = 0.5$ no shows.
      (c) $(6)(0.10) = 0.6$ seconds.

6.75 $\mu = 0(0.295) + 1(0.491) + 2(0.196) + 3(0.018) = 0.937$ and $\mu = \frac{3 \cdot 5}{5 + 11} = \frac{15}{16} = 0.9375$.

6.77   Practice exercise.

6.79   (a)   At least $1 - \dfrac{1}{4^2} = \dfrac{15}{16} \approx 0.94$;

(b)   At least $1 - \dfrac{1}{2.5^2} = 1 - \dfrac{1}{6.25} = 0.84$;

(c)   At least $1 - \dfrac{1}{2.8^2} \approx 0.87$.

6.81   The probability of getting a value between $\mu - 2\sigma$ and $\mu + 2\sigma$ is at least $1 - \dfrac{1}{2^2} = \dfrac{3}{4}$; and of getting a value between $\mu - 3\sigma$ and $\mu + 3\sigma$ is at least $1 - \dfrac{1}{3^2} = \dfrac{8}{9}$. Thus:

(a)   Since $50 - 2\sigma = 50 - 2 \cdot 5 = 40$ and $50 + 2\sigma = 50 + 2 \cdot 5 = 60$, we can assert with a probability of at least $\dfrac{3}{4}$ that the restaurant will serve between 40 and 60 meals in an evening; and

(b)   Since $50 - 3\sigma = 50 - 15 = 35$, and $50 + 3\sigma = 50 + 15 = 65$, we can assert with a probability of at least $\dfrac{8}{9}$ that the restaurant will serve between 35 and 65 meals in an evening.

Chapter 7

The Normal Distribution

7.1   Practice exercise.

7.3   Practice exercise.

7.5   Practice exercise.

7.7   (a)   Process B is preferable;   (b)   no "right" answer.

7.9   (a)   0.3023;
     (b) $0.5000 + 0.2422 = 0.7422$;
     (c) $0.5000 - 0.4778 = 0.0222$;
     (d) $0.3643 - 0.3413 = 0.0230$.

7.11   Practice exercise.

7.13   (a)   2.00;
      (b)   1.11;
      (c)   0.55;
      (d)   3.08 or 3.09.

7.15   (a) $\dfrac{70.00 - 56.4}{4.8} = 2.83$, and the entry corresponding to $z = 2.83$ in Table II is 0.4977.
      $0.5000 + 0.4977 = 0.9977$.

      (b) $\dfrac{50.0 - 56.4}{4.8} = -1.33$ and the entry corresponding to $z = -1.33$ in Table II is 0.4082.
      $0.5000 - 0.4082 = 0.0918$.

      (c)   Since $z = \dfrac{50.0 - 56.4}{4.8} = -1.33$ and $z = \dfrac{70 - 56.4}{4.8} = 2.83$ and the corresponding entry to
      $z = 2.83$ in Table II is 0.4977, the probability is $0.4082 + 0.4977 = 0.9059$.

7.17   (a)   Since $0.5000 - 0.0250 = 0.4750$ corresponds to $z = 1.96$ in Table II, we get $z_{0.025} = 1.96$.
      (b)   Since $0.5000 - 0.0050 = 0.4950$, and 0.4949 and 0.4951 are the entries in Table II
      corresponding to $z = 2.57$ and $z = 2.58$, we get $z_{0.005} = \dfrac{2.57 + 2.58}{2} = 2.575$.

7.19   (See figure below.) The entry in Table II nearest to $0.500 - 0.1000 = 0.4000$ is 0.3997 corresponding to $z = 1.28$. Thus, $\dfrac{100 - 74.4}{\sigma} = 1.28$ and it follows that $25.6 = 1.28\sigma$.

$\sigma = \dfrac{25.6}{1.28} = 20.$

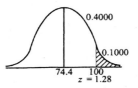

7.21   The entry of Table II of $0.5000 - 0.3300 = 0.1700$ corresponds to $z = 0.44$. Thus $\dfrac{245 - \mu}{\sigma} = -0.44$, and it follows that $\mu = 0.44\sigma + 245$. The entry in Table II of $0.5000 - 0.4800 = 0.0200$ corresponds to $z = 0.05$. Thus, $\dfrac{260 - \mu}{\sigma} = 0.05$, and it follows that $\mu + 0.05\sigma = 260$. The two equations can be solved for $\mu \approx 258.47$ and $\sigma \approx 30.61$.

7.23   (a) $z = \dfrac{49.00 - 50.25}{0.63} \approx -1.98$, and the value of z from Table II is 0.4761, and $.5000 - 0.4761 = 0.0239$.

   (b) $z = \dfrac{50.5 - 50.25}{0.63} \approx 0.40$ and the value of z from Table II is 0.1554.

   (c) $z = \dfrac{51.0 - 50.25}{0.63} \approx -0.40$ and the value of z from Table II is 0.1554.

   $z = \dfrac{50.0 - 50.25}{0.63} \approx 1.19$ and the value of z from Table II is 0.3830 and $0.1554 + 0.3830 = 0.5384$.

7.25   The continuity correction requires that we use the number 50.5 instead of 50. Substituting 50.5 into the formula $z = \dfrac{x - \mu}{\sigma}$, we get $z = \dfrac{50.5 - 43.3}{4.6} = 1.57$ and the value of z from Table II is 0.4418. Finally, $0.5000 - 0.4418 = 0.0582$.

7.27   (See figure below) Since the entry in Table II nearest to $0.5000 - 0.1500 = 0.3500$ is 0.3508 corresponding to $z = 1.04$, we get $\dfrac{x - 4.64}{0.25} = 1.04$. It follows that $x = 4.64 + 1.04(0.25)$ and $x = 4.64 + 0.26 = 4.90$ inches.

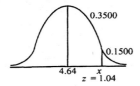

7.29   Practice exercise.

7.31   Since $z = \dfrac{19.5 - 15.3}{3.4} \approx 1.24$ and $z = \dfrac{20.5 - 15.3}{3.4} \approx 1.53$, and the corresponding entries in Table II are 0.3925 and 0.4370, the probability is $0.4370 - 0.3925 = 0.0445$.

7.33 $z = \dfrac{4.5 - 7.1}{3.2} \approx -0.81$ and $z = \dfrac{10.5 - 7.1}{3.2} \approx 1.06$ and the entries corresponding to $z = 0.81$ and $z = 1.06$ in Table II are 0.2910 and 0.3554, so that the probability is $0.2910 + 0.3554 = 0.6464$, and about $(1,000)(0.6464) \approx 646$ patients can expect to be hospitalized from five to ten days.

7.35   (a)   The conditions are satisfied. $np = 10.8$ and $n(1 - p) = 7.2$.
(b)   The conditions are not satisfied. $np = 4.5$
(c)   The conditions are not satisfied. $np = 5$ and $n(1 - p) = 5$.

7.37 $\mu = 24(0.62) = 14.88$ and $\sigma = \sqrt{24(0.62)(0.38)} \approx 2.38$.
Since $z = \dfrac{14.5 - 14.88}{2.38} \approx -0.16$ and $z = \dfrac{15.5 - 14.88}{2.38} \approx 0.26$, and the entries in Table II corresponding to $z = -0.16$ and $z = 0.26$ are 0.0636 and 0.1026, the probability is $0.0636 + 0.1206 = 0.1662$. Therefore, the error of the approximation is $0.1662 - 0.1661 = 0.0001$.

7.39   (a) $z = \dfrac{30.5 - 34.9}{4.8} = -0.92$ and $z = \dfrac{31.5 - 34.9}{4.8} = -0.7083$; the corresponding values for z are 0.3212 and 0.2611 and the probability is $0.3212 - 0.2611 = 0.0601$.

(b) $z = -0.71$; the probability is $.5000 - 0.2611 = 0.2389$.

7.41 $\mu = np = 500(0.06) = 30$ and $\sigma = \sqrt{np(1 - p)} = \sqrt{500(0.06)(0.94)} = 5.31$
$z = \dfrac{24.5 - 30}{5.31} = -1.04$. The area corresponding to $z = -1.04$ is 0.3508 and the probability is $0.5000 - 0.3508 = 0.1492$.

7.43   (a) $\mu = (100)(0.50) = 50$, and $\sigma = \sqrt{(100)(0.50)(0.50)} = 5$.
Since $z = \dfrac{48.5 - 50}{5} = -0.3$ and $z = \dfrac{51.5 - 50}{5} = 0.3$, and the entry corresponding to 0.3 in Table II is 0.1179, the probability is $0.1179 + 0.1179 = 0.2358$.
(b) $\mu = (1,000)(0.50) = 500$, and $\sigma = \sqrt{(1000)(0.50)(0.50)} \approx 15.81$.
Since $z = \dfrac{489.5 - 500}{15.81} \approx -0.66$ and $z = \dfrac{510.5 - 500}{15.81} \approx 0.66$, and the entry corresponding to 0.66 in Table II is 0.2454, the probability is $0.2454 + 0.2454 = 0.4908$.
(c) $\mu = (10,000)(0.50) = 5,000$, and $\sigma = \sqrt{(10,000)(0.50)(0.50)} = 50.0$.
Since $z = \dfrac{4,899.5 - 5,000}{50} = -2.01$ and $z = \dfrac{5,100.5 - 5,000}{50} = 2.01$, and the entry corresponding to 2.01 in Table II is 0.4778, the probability is $0.4778 + 0.4778 = 0.9556$.

7.45   (a)   The error is $0.0767 - 0.0823 = -0.0056$.

        (b)   The error is $0.1999 - 0.2033 = -0.0034$.

7.47   We approximate the distribution as normal with mean $\mu = 45$ and standard deviation $\sigma = \sqrt{45} \approx 6.708$.

        (a)   We seek the area to the right of 49.5. The corresponding z is $\dfrac{49.5 - 45}{6.708} \approx 0.67$, and the probability is $0.5000 - 0.2486 = 0.2514$.

        (b)   We seek the area to the left of 44.5. The corresponding z is $\dfrac{44.5 - 45}{6.708} \approx -0.075$, and the probability is $0.5000 - \dfrac{1}{2}(0.0279 + 0.0319) = 0.5000 - 0.0299 = 0.4701$.

Chapter 8

Sampling and Sampling Distributions

8.1   Practice exercise.

8.3   (a) $\binom{4}{3} = 4,$ (b) $\binom{6}{3} = 20,$

(c) $\binom{12}{3} = 220,$ (d) $\binom{300}{3} = 4,455,100.$

8.5   (a) $\dfrac{1}{\binom{4}{3}} = \dfrac{1}{4} = 0.2500$ (b) $\dfrac{1}{\binom{6}{3}} = \dfrac{1}{20} = 0.0500$

(c) $\dfrac{1}{\binom{12}{3}} = \dfrac{1}{220} = 0.0045$ (d) $\dfrac{1}{\binom{18}{3}} = \dfrac{1}{816} = 0.0012$

8.7   abc, bcd, abd, acd.

8.9   (a) M A U T, A U T O, U T O M, T O M A, O M A U,
        (b) 1/5.
        (c) 4/5.

8.11   03, 11, 22, 13, 01, 08, 24, 21, 10, 17, 04, 09.

8.13   Students 2997, 1487, 3852, 0372, 2705, 1005, 3113, 0732, 3780, 3815, 2742, 1764, 3247, 3375, and 3700.

8.15   (a) The probability that any individual will be in the sample is $\dfrac{10}{60} = \dfrac{1}{6}.$ You can also do this formally as $\dfrac{\left(\begin{array}{c}\text{number of}\\\text{samples}\\\text{with Randy}\end{array}\right)}{\left(\begin{array}{c}\text{number of}\\\text{samples}\\\text{overall}\end{array}\right)} = \dfrac{\binom{59}{9}}{\binom{60}{10}}$ and this will reduce to $\dfrac{1}{6}.$ The number of samples with Randy is obtained by finding the number of ways to select the remaining 9 persons from the other 59.

(b) The probability that Susan will be in the sample is $\dfrac{1}{6},$ exactly as in part (a).

(c)   The probability that both will be in the sample is

$$\frac{\left(\begin{array}{c}\text{number of}\\\text{samples}\\\text{with both}\end{array}\right)}{\left(\begin{array}{c}\text{number of}\\\text{samples}\\\text{overall}\end{array}\right)} = \frac{\binom{58}{8}}{\binom{60}{10}} = \frac{10\cdot9}{60\cdot59} = \frac{3}{118} \approx 0.0254.$$

(d) $\frac{3}{118} < \frac{1}{36}$ since $3 \cdot 36 = 108 < 118$.

8.17   Practice exercise.

8.19   (a) $\mu = \dfrac{2+4+6+8}{4} = 5.$

$$\sigma^2 = \frac{1}{4}\left((2-5)^2 + (4-5)^2 + (6-5)^2 + (8-5)^2\right)$$

$$= \frac{1}{4}\left((-3)^2 + (-1)^2 + 1^2 + 3^2\right)$$

$$= \frac{20}{4} = 5. \text{ Then } \sigma = \sqrt{5} \approx 2.236.$$

(b)   The sixteen possible samples are 2 and 2, 2 and 4, 2 and 6, 2 and 8, 4 and 2, 4 and 4, 4 and 6, 4 and 8, 6 and 2, 6 and 4, 6 and 6, 6 and 8, 8 and 2, 8 and 4, 8 and 6, and 8 and 8. Their means are 2, 3, 4, 5, 3, 4, 5, 6, 4, 5, 6, 7, 5, 6, 7, and 8.

(c)

Sample Mean	Probability
2	$\frac{1}{16}$
3	$\frac{2}{16}$
4	$\frac{3}{16}$
5	$\frac{4}{16}$
6	$\frac{3}{16}$
7	$\frac{2}{16}$
8	$\frac{1}{16}$

(d) The mean of the sampling distribution is five and the variance of the sampling distribution is

$$(2-5)^2 \cdot \frac{1}{16} + (3-5)^2 \cdot \frac{2}{16} + (4-5)^2 \cdot \frac{3}{16} + (5-5)^2 \cdot \frac{4}{16} + (6-5)^2 \cdot \frac{3}{16} + (7-5)^2 \cdot \frac{2}{16} + (8-5)^2 \cdot \frac{1}{16}$$

$$= \frac{1}{16}(9+8+3+0+3+8+9) = \frac{40}{16} = 2.5. \text{ The standard deviation is } \sqrt{2.5} \approx 1.581.$$

Since the sampling is with replacement, the standard deviation formula is

$$\frac{\sigma}{\sqrt{n}} = \frac{\sqrt{5}}{\sqrt{2}} = \sqrt{2.5} \approx 1.581. \text{ Again the calculations agree.}$$

8.21 (a) The ratio of the two standard errors is $\dfrac{\dfrac{\sigma}{\sqrt{27}}}{\dfrac{\sigma}{\sqrt{3}}} = \dfrac{\sqrt{3}}{\sqrt{27}} = \dfrac{1}{3}$.

(b) The ratio of the two standard errors is $\dfrac{\dfrac{\sigma}{\sqrt{3}}}{\dfrac{\sigma}{\sqrt{27}}} = \dfrac{\sqrt{27}}{\sqrt{3}} = \dfrac{3}{1}$.

(c) The ratio of the two standard errors is $\dfrac{\dfrac{\sigma}{\sqrt{243}}}{\dfrac{\sigma}{\sqrt{27}}} = \dfrac{\sqrt{27}}{\sqrt{243}} = \dfrac{1}{3}$.

(d) The ratio of the two standard errors is $\dfrac{\dfrac{\sigma}{\sqrt{27}}}{\dfrac{\sigma}{\sqrt{243}}} = \dfrac{\sqrt{243}}{\sqrt{27}} = \dfrac{3}{1}$.

8.23 The finite population correction factor is $\sqrt{\dfrac{10,000-100}{10,000-1}} = 0.995$.

If the correction factor had been 1, the answer to a standard error of the mean problem would be unchanged. Since 0.995 is very close to 1, the answer would be almost unchanged, and would probably not be used.

8.25 (a) 0.889 or 0.866;
(b) 0.871 or 0.866;
(c) 0.975 or 0.975;
(d) 0.912 or 0.910.

8.27 Computer exercise.

8.29 Computer exercise.

8.31   Substituting $\sigma = 9$ and $n = 144$ into the formula $\sigma_{\bar{x}} = \dfrac{\sigma}{\sqrt{n}}$, we get $\sigma_{\bar{x}} = \dfrac{9}{\sqrt{144}} = \dfrac{9}{12} = 0.75$,

and it follows that we can assert with the probability $1 - \dfrac{1}{3^2} = \dfrac{8}{9} \approx 0.89$ that the error will be

less than $k \cdot \sigma_{\bar{x}} = (3)(0.75) = 2.25$.

8.33   (a)   Since $n = 36$ and $\sigma = 11$, the sampling distribution of the mean has the standard

deviation $\dfrac{11}{\sqrt{36}} = \dfrac{11}{6}$. Substituting $\dfrac{11}{6}$ for $\sigma_{\bar{x}}$ in $k \cdot \sigma_{\bar{x}} = 3.3$, we get $k = 1.80$, so that

the probability is at least $1 - \dfrac{1}{(1.80)^2} \approx 0.691$.

   (b)   The probability is given by the area under the standard normal curve between $z = -1.80$
and $z = 1.80$. The entry corresponding to $z = 1.80$ in Table II is 0.4641, so that the
probability is $2(0.4641) = 0.9282$.

8.35 $\sigma_{\bar{x}} = \dfrac{2}{\sqrt{100}} = \dfrac{2}{10} = 0.20$, $k = \dfrac{0.25}{0.20} = 1.25$ and the probability is at least $1 - \dfrac{1}{1.25^2} = 0.36$.

8.37   The standard deviation of the mean for a sample size of 49 is $\dfrac{250}{\sqrt{49}} = \dfrac{250}{7} \approx 35.714$ lines.

   (a)   Since $z = \dfrac{50}{35.714} \approx 1.40$ and the corresponding entry in Table II is 0.4192, the
probability is $2(0.4192) = 0.8384$.

   (b)   Since $z = \dfrac{25}{35.714} \approx 0.70$ and the corresponding entry in Table II is 0.2580, the
probability is $2(0.2580) = 0.5160$.

8.39   Practice exercise.

8.41   Since $1.25 \cdot \dfrac{\sigma}{\sqrt{n}} = \dfrac{\sigma}{\sqrt{1,600}}$, it follows that $\sqrt{n} = 1.25 \cdot \sqrt{1,600} = 1.25 \cdot 40 = 50$ and

$n = 50^2 = 2,500$.

R.83 (a) 0.0467

(b) $0.3110 + 0.2765 + 0.1382 + 0.0369 + 0.0041 = 0.7667$

(c) $0.0467 + 0.1866 = 0.2333$

R.85 (a) $z = \dfrac{15.0 - 18.0}{3.2} = -0.94$ and the probability is $0.5000 - 0.3264 = 0.1736$.

(b) $z = \dfrac{20.0 - 18.0}{3.2} = 0.62$, and the corresponding area is 0.2324. On the other side of the mean we have $\dfrac{15.0 - 18.0}{3.2} = -0.94$ and the corresponding area is 0.3264. Summing the areas on both sides of the mean, we have the probability $0.3264 + 0.2324 = 0.5588$.

(c) $z = 0.62$ and the corresponding area is 0.2324. Thus the probability is $0.5000 - 0.2324 = 0.2676$.

R.87 (a) No. In practice, the finite population factor is omitted unless the sample, n, is at least 5% of the population, N. In this exercise, $\dfrac{20}{1{,}000} = 0.02$, or 2%, which is less than 5%.

R.89 (a) 0.003;

(b) $0.092 + 0.042 + 0.014 + 0.003 = 0.151$;

(c) 0.153;

(d) $0.153 + 0.196 = 0.349$.

R.91 (a) The probability is $\frac{1}{2}(0.5)(0.5) = 0.125$.

(b) The probability is $\frac{1}{2}(0.8)(0.8) - \frac{1}{2}(0.4)(0.4) = 0.32 - 0.08 = 0.24$.

R.93 22, 13, 01.

R.95 (a) $f(0) = e^{-3.5} \cdot \dfrac{(3.5)^0}{0!} \approx 0.030$;

(b) $f(2) = e^{-3.5} \cdot \dfrac{(3.5)^2}{2!} = e^{-3.5} \cdot (6.125) \approx (0.030)(6.125) \approx 0.185$;

(c) $f(4) = e^{-3.5} \cdot \dfrac{(3.5)^4}{4!} = e^{-3.5} \cdot (6.253) \approx (0.030)(6.253) \approx 0.189$;

(d) $f(6) = e^{-3.5} \cdot \dfrac{(3.5)^6}{6!} = e^{-3.5} \cdot (2.553) \approx (0.030)(2.553) \approx 0.077$.

R.97   (a) $\mu = 0(0.410) + 1(0.410) + 2(0.154) + 3(0.026) + 4(0.002) = 0.804$, or approximately 0.8;

and $\sigma^2 = (0 - 0.8)^2(0.410) + (1 - 0.8)^2(0.410) + (2 - 0.8)^2(0.154) + (3 - 0.8)^2(0.026)$

$+ (4 - 0.8)^2(0.002) \approx 0.65$.

(b) $\mu = n \cdot p = 4(0.20) = 0.8$;

$\sigma^2 = n \cdot p(1 - p) = (4)(0.20)(0.80) = 0.64$.

R.99 $z = \dfrac{x - \mu}{\sigma} = \dfrac{30 - 36}{6} = -1$, and the corresponding area is 0.3413; and the probability that the

rope will last at least 30 months is $0.5000 + 0.3413 = 0.8413$.

R.101   (a) $\dfrac{\binom{8}{2}\binom{6}{2}}{\binom{14}{4}} = \dfrac{28 \cdot 15}{1{,}001} \approx 0.420$;

(b) $\dfrac{\binom{8}{1}\binom{6}{3}}{\binom{14}{4}} = \dfrac{8 \cdot 20}{1{,}001} \approx 0.160$.

R.103   (a)   Since the sum of the values is $0.98 < 1$, they cannot be the values of a probability distribution.

(b)   Since $f(4)$ is negative, the given values cannot be the values of a probability distribution.

(c)   Since the values are all non-negative and less than 1, and their sum is equal to 1, they can be the values of a probability distribution.

R.105 $\mu = 500(0.73) = 365$ cars and $\sigma = \sqrt{500(0.73)(0.27)} \approx 9.927$.

Since $z = \dfrac{350.5 - 365}{9.927} \approx -1.46$ and the entry in Table II corresponding to $z = 1.46$ is 0.4279,

the probability is $0.5000 - 0.4279 = 0.0721$.

R.107   (a)   Since $np = 50 \cdot \dfrac{1}{5} = 10 > 5$ and $n(1 - p) = 50 \cdot \dfrac{4}{5} = 40 > 5$, the conditions are satisfied.

(b)   Since $np = 80 \cdot \dfrac{1}{40} = 2 < 5$, the conditions are not satisfied.

(c)   Since $n(1 - p) = 100(0.04) = 4 < 5$, the conditions are not satisfied.

(d)   Since $np = 200 \cdot \dfrac{24}{25} = 192 > 5$ and $n(1 - p) = 200 \cdot \dfrac{1}{25} = 8 > 5$, the conditions are satisfied.

R.109 $\sqrt{\dfrac{N - n}{N - 1}} = \sqrt{\dfrac{100 - 50}{100 - 1}} \approx 0.711$

R.111   (a)   Since $n = 18$ is greater than $0.05(150 + 150) = 15$, the condition is not satisfied;

(b)   since $n = 15$ is less than $0.05(100 + 300) = 20$, the condition is satisfied;

(c)   since $n = 50$ is greater than $0.05(400 + 200) = 30$, the condition is not satisfied.

R.113 Substituting $\sigma = 25$ and $n = 81$ into the formula for the standard error of the mean, $\sigma_{\bar{x}} = \dfrac{\sigma}{\sqrt{n}}$

we get $\sigma_{\bar{x}} = \dfrac{25}{\sqrt{81}} = 2.8$, and it follows that we can assert with a probability of at least

$1 - \dfrac{1}{3^2} = \dfrac{8}{9} \approx 0.89$ that the error is less than $k \cdot \sigma_{\bar{x}} = 3(2.8) = 8.4$.

R.115 (a) Since $n = 156 \geq 100$ and $np = 156 \cdot \dfrac{1}{26} = 6 < 10$, the conditions for the Poisson approximation are satisfied.

     (b) Since $np = 156 \cdot \dfrac{1}{26} = 6 > 5$ and $156 \cdot \dfrac{25}{26} = 150 > 5$, the conditions for the normal approximation are satisfied.

R.117 Since $n \cdot p = 250(0.018) = 4.5$, the probability that exactly five will not receive their

newspapers is $e^{-4.5} \cdot \left(\dfrac{4.5^5}{5!} \right) = 0.011 \cdot \dfrac{1,845.2813}{120} \approx 0.1692$.

Chapter 9

Problems of Estimation

9.1 Practice exercise.

9.3 The maximum error is $E = 2.575 \cdot \dfrac{0.020}{\sqrt{42}} \approx 0.008$ inches.

9.5 The maximum error is $E = 2.575 \cdot \dfrac{0.50}{\sqrt{100}} \approx 0.1288$ minutes.

9.7 The maximum error is $E = 2.575 \cdot \dfrac{2.9}{\sqrt{56}} \approx 0.9979$ chair, or 1.0 chair.

9.9 (a) The maximum error is $E = 1.96 \cdot \dfrac{0.5}{\sqrt{64}} = 0.1225$ ounces of honey.

  (b) The maximum error is $E = 2.575 \cdot \dfrac{0.5}{\sqrt{64}} \approx 0.1609$ ounces of honey.

9.11 The maximum error is $E = 2.575 \cdot \dfrac{5.5}{\sqrt{200}} \approx 1$ minute.

9.13 $n = \left(\dfrac{1.96 \cdot 15}{3.0} \right)^2 \approx 96.04$, which we move up to the next integer, 97 children.

9.15 $n = \left(\dfrac{2.575 \cdot 7.1}{3} \right)^2 \approx 37.14$, which we move up to the next integer, 38 cars.

9.17 We require that n be at least $\left(z_{0.025} \cdot \dfrac{\sigma}{0.30\sigma} \right)^2 = \left(\dfrac{1.96}{0.30} \right)^2 \approx 42.68$. We should use $n = 43$.

9.19 (a) $25.1 - 1.96 \cdot \dfrac{8.5}{\sqrt{49}} < \mu < 25.1 + 1.96 \cdot \dfrac{8.5}{\sqrt{49}}$, so that 22.72 tons $< \mu <$ 27.48 tons.

  (b) $25.1 - 2.575 \cdot \dfrac{8.5}{\sqrt{49}} < \mu < 25.1 + 2.575 \cdot \dfrac{8.5}{\sqrt{49}}$, so that 21.97 tons $< \mu <$ 28.23 tons.

9.21 $6{,}300 - 2.821 \cdot \dfrac{500}{\sqrt{10}} < \mu < 6{,}300 + 2.821 \cdot \dfrac{500}{\sqrt{10}}$ so that approximately
5,853.96 pounds $< \mu <$ 6,746.04 pounds.

9.23 $85 - 1.96 \cdot \dfrac{9.2}{\sqrt{90}} < \mu < 85 + 1.96 \cdot \dfrac{9.2}{\sqrt{90}}$, so that approximately $83.10 < \mu < 86.90$ blood tests.

9.25   (a) $11 - 1 = 10$, and under column headed 0.05 is 1.812,
         (b) $12 - 1 = 11$, and under column headed 0.005 is 3.106,
         (c) $13 - 1 = 12$, and under column headed 0.025 is 2.179,
         (d) $14 - 1 = 13$, and under column headed 0.010 is 2.650.

9.27   Substituting $n = 10$, $s = 1.1$, and $t_{0.005} = 3.250$ (the entry in Table III for 9 degrees of freedom)
we get $E = 3.250 \cdot \dfrac{1.1}{\sqrt{10}} \approx 1.13$ pounds.

9.29 $\bar{x} = \dfrac{9 + 14 + 7 + 8 + 11 + 5}{6} = \dfrac{54}{6} = 9$, $s^2 = \dfrac{0^2 + 5^2 + (-2)^2 + (-1)^2 + 2^2 + (-4)^2}{6-1} = \dfrac{50}{5} = 10$, and

$s = \sqrt{10} \approx 3.16$. The interval is $9 \pm 2.571 \cdot \dfrac{3.16}{\sqrt{6}}$, 9 ± 3.32, so that $5.68 < \mu < 12.32$.

9.31 $\bar{x} = \dfrac{18 + 19 + 23 + 19 + 21 + 20}{6} = 20$ minutes, and

$s^2 = \dfrac{(-2)^2 + (-1)^2 + 3^2 + (-1)^2 + 1^2 + 0^2}{6-1} = 3.2$ minutes so that $s = \sqrt{3.2} \approx 1.789$. The

confidence interval is $20 \pm 2.571 \cdot \dfrac{1.789}{\sqrt{6}} = 20 \pm 1.9$, and $18.1 < \mu < 21.9$ minutes.

9.33   (a)   for $n = 10$ (9 df), the ratio is $\dfrac{3.250}{2.262} \approx 1.44$;

         (b)   for $n = 20$ (19 df), the ratio is $\dfrac{2.861}{2.093} \approx 1.37$;

         (c)   for $n = 30$ (29 df), the ratio is $\dfrac{2.756}{2.045} \approx 1.35$.

9.35   Computer exercise.

9.37   Practice exercise.

9.39 $\dfrac{5.15}{1 + \dfrac{1.96}{\sqrt{64}}} < \sigma < \dfrac{5.15}{1 - \dfrac{1.96}{\sqrt{64}}}$, so that $4.14 < \sigma < 6.82$

9.41 $\dfrac{1.84}{1 + \dfrac{2.575}{\sqrt{240}}} < \sigma < \dfrac{1.84}{1 - \dfrac{2.575}{\sqrt{240}}}$, so that 1.58 minutes $< \sigma < 2.21$ minutes.

9.43   Practice exercise.

9.45   Since $\dfrac{x}{n} = \dfrac{30}{100} = 0,30$ and $z_{0.025} = 1.96$ we get

$$0.30 - 1.96\sqrt{\frac{(0.30)(0.70)}{100}} < p < 0.30 + 1.96\sqrt{\frac{(0.30)(0.70)}{100}} \text{ , and hence } 0.21 < p < 0.39.$$

9.47   Since $z_{0.005} = 2.575$ the maximum error is $E = 2.575\sqrt{\dfrac{(0.15)(0.85)}{500}} = 0.041$.

9.49   Since $\dfrac{x}{n} = \dfrac{1,030}{2,500} \cdot 100 \approx 41.2$ and $z_{0.025} = 1.96$, the maximum error is

$$E = 1.96\sqrt{\frac{(41.2)(58.8)}{2,500}} = 1.93\% \text{ .}$$

9.51   Since $\dfrac{x}{n} = \dfrac{900}{2,500} = 0.36$ and $z_{0.025} = 1.96$, the maximum error is $E = 1.96\sqrt{\dfrac{(0.36)(0.64)}{2,500}} =$

0.0188, and rounding to two decimals, we get 0.02.

9.53   Practice exercise.

9.55 $n = (0.30)(0.70)\left(\dfrac{2.575}{0.02}\right)^2 \approx 3,482$ rounded up to the nearest integer.

9.57   Since 0.20 is closer to $\dfrac{1}{2}$ than 0.10, we get $n = (0.20)(0.80)\cdot\left(\dfrac{1.96}{0.05}\right)^2$

$= (0.16)(1,536.64) \approx 246$, rounded up to the nearest integer.

10.1   Practice exercise.

10.3   (a)   We would commit a Type I error if we erroneously reject the hypothesis that the average number of gallons of gasoline purchased is 9.3.

       (b)   We would commit a Type II error if we erroneously accept null hypothesis that the number of gallons of gasoline purchased is 9.3.

10.5   Test the hypothesis that the use of telemarketing is not effective in increasing automobile sales.

10.7   (a)   Since $z = \dfrac{13.7 - 13.4}{\dfrac{0.8}{\sqrt{40}}} \approx 2.37$ and the corresponding entry in Table II is 0.4911, the

probability of a Type I error is $2(0.5000 - 0.4911) = 2(0.0089) = 0.0178 \approx 0.02$.

       (b)   Since $z = \dfrac{13.1 - 13.7}{\dfrac{0.8}{\sqrt{40}}} \approx -4.74$, and the value in Table II closest to $z = 4.74$ is $z = 5.00$,

the probability of a Type II error is 0.5000.

10.9   (a)   Since $z = \dfrac{13.2 - 13.5}{\dfrac{0.8}{\sqrt{40}}} \approx -2.37$ and $z = \dfrac{13.6 - 13.5}{\dfrac{0.8}{\sqrt{40}}} \approx 0.79$, and the normal curve areas

corresponding to $z = 2.37$ and $z = 0.79$ are 0.4911 and 0.2852, the probability of a Type II error when μ is really 13.5 is $0.4911 + 0.2852 = 0.7763 \approx 0.78$. Since

$z = \dfrac{13.2 - 13.3}{\dfrac{0.8}{\sqrt{40}}} \approx -0.79$ and $z = \dfrac{13.6 - 13.3}{\dfrac{0.8}{\sqrt{40}}} \approx 2.37$, the probability of a Type II error

when μ is really 13.3 is also $0.7763 \approx 0.78$.

       (b)   Since $z = \dfrac{13.2 - 13.6}{\dfrac{0.8}{\sqrt{40}}} \approx -3.16$ and $z = \dfrac{13.6 - 13.6}{\dfrac{0.8}{\sqrt{40}}} \approx 0$, and the area under the curve

to the left of $z = -3.16$ is negligible, the probability of a Type II error when μ is really 13.6 is 0.5000. By symmetry the probability of a Type II error when μ is really 13.2 is also 0.5000.

(c)   Since $z = \dfrac{13.2 - 13.8}{\frac{0.8}{\sqrt{40}}} \approx -4.74$ and $z = \dfrac{13.6 - 13.8}{\frac{0.8}{\sqrt{40}}} \approx -1.58$, and the area under the

curve to the left of $z = -4.74$ is negligible, while the entry in Table II corresponding to $z = 1.58$ is 0.4429, the probability of a Type II error when μ is really 13.8 is 0.5000 − 0.4429 = 0.0571 ≈ 0.06. By symmetry the probability of a Type II error when μ is really 13.0 is also 0.06.

10.11   Practice exercise.

10.13   (a)   1.   <u>Hypotheses</u>
$$H_0 : \mu = 15.0$$
$$H_A : \mu \neq 15.0$$
If the investigation was made for the purpose of comparing the length of time required to assemble a chair to a predetermined norm, we use the two-sided test as directed.

2.   <u>Level of significance</u>
$$\alpha = 0.05$$

3.   <u>Criterion</u>
Reject H_0 if $z \leq -1.96$ or $z \geq 1.96$

4.   <u>Calculations</u>
$$z = \frac{15.4 - 15.0}{\frac{2.4}{\sqrt{150}}} \approx 2.04$$

5.   <u>Decision</u>
Since 2.04 ≥ 1.96, we must reject the null hypothesis. Accept the alternative hypothesis that μ does not equal 15.

(b)   1.   <u>Hypothesis</u>
Same as above

2.   <u>Level of significance</u>
$$\alpha = 0.01$$

3.   <u>Criterion</u>
Reject H_0 if $z \leq -2.575$ or $z \geq 2.575$

4.   <u>Calculations</u>
$$z = \frac{15.4 - 15.0}{\frac{2.4}{\sqrt{150}}} \approx 2.04$$

5.   <u>Decision</u>
Since $z = 2.04$ falls between −2.575 and 2.575, we cannot reject the null hypothesis.

10.15   1.   <u>Hypotheses</u>
$$H_0 : \mu = 24.0$$
$$H_A : \mu \neq 24.0$$

2.   <u>Level of significance</u>
$$\alpha = 0.05$$

3. Criterion
    Reject the null hypothesis if $z < -1.96$ or $z > 1.96$.

4. Calculations
$$z = \frac{23 - 24}{\frac{7}{\sqrt{36}}} \approx -0.86$$

5. Decision
    Since -0.86 lies between -1.96 and 1.96 the null hypothesis cannot be rejected.

10.17 1. Hypotheses
$$H_0 : \mu = 8.0$$
$$H_A : \mu > 8.0$$

2. Level of significance
$$\alpha = 0.01$$

3. Criterion
    Reject H_0 if $z > 2.33$.

4. Calculations
$$z = \frac{8.2 - 8.0}{\frac{0.4}{\sqrt{49}}} = 3.50$$

5. Decision
    Since $z = 3.50$ is greater than 2.33, the null hypothesis must be rejected. Accept the alternative hypothesis that $\mu > 8.0$.

10.19 (a) $z = \dfrac{8.62 - 8.5}{\frac{0.55}{\sqrt{60}}} \approx 1.69$

(b) The criterion (step 3 of the hypothesis testing process) is to reject H_0 if $z \le -1.96$ or $z \ge 1.96$. With $z = 1.69$, the null hypothesis cannot be rejected, thus concluding that the running time has not significantly changed.

(c) The criterion (step 3 of the hypothesis testing process) is to reject H_0 if $z \ge 1.645$. With $z = 1.69$, the engineer will reject the null hypothesis, concluding that the new process has significantly improved the running time.

(d) The criterion (step 3 of the hypothesis testing process) is to reject H_0 if $z \le -1.645$. With $z = 1.69$, the second engineer cannot reject the null hypothesis and conclude that there is no evidence that the running time has worsened.

(e) If you have no explanation about the purpose of the new process, you must use the two-sided alternative H_A: $\mu \ne 8.5$. Without a level of significance, you can use the p-value. Since the value in Table II corresponding to $z = 1.69$ is 0.4545, the p-value is $2(0.5 - 0.4545) = 0.0910$.

10.21 Practice exercise.

10.23   The sample size $n = 10$ is small, and we must begin by stating that the work rests on the assumption that the values are sampled from a population which has roughly the shape of a normal distribution.

     1.   <u>Hypotheses</u>

$$H_0 : \mu = 10.0$$
$$H_A : \mu < 10.0$$

     2.   <u>Level of significance</u>

$$\alpha = 0.05$$

     3.   <u>Criterion</u>
        Reject the null hypothesis where $t \le -1.833$, where 1.833 is the value of $t_{0.05}$ for $10 - 1 = 9$ degrees of freedom.

     4.   <u>Calculations</u>
        Solving for s using the 10 items of the sample we get $\bar{x} = 9$ and $s = 1.56$, and

$$t = \frac{9 - 10}{\frac{1.56}{\sqrt{10}}} \approx -2.03$$

     5.   <u>Decision</u>
        Since $z = -2.03$ is less than -1.833, the null hypothesis must be rejected.

10.25   The sample size $n = 10$ is small, and we must begin by stating that the work rests on the assumption that the values are sampled from a population which has roughly the shape of a normal distribution.

     1.   <u>Hypotheses</u>

$$H_0 : \mu = 0.7500 \text{ inch}$$
$$H_A : \mu \ne 0.7500 \text{ inch}$$

     2.   <u>Level of significance</u>

$$\alpha = 0.01$$

     3.   <u>Criterion</u>
        Reject the null hypothesis if $t \le -3.250$ or $t \ge 3.250$; where 3.250 is the value of $t_{0.005}$ for $10 - 1 = 9$ degrees of freedom; otherwise, accept H_0 or reserve judgement.

     4.   <u>Calculations</u>

$$t = \frac{0.7510 - 0.7500}{\frac{0.0030}{\sqrt{10}}} \approx 1.05$$

     5.   <u>Decision</u>
        Since $t = 1.05$ falls between -3.250 and 3.250, the null hypothesis cannot be rejected.

10.27   Begin by noting that the sample size $n = 8$ is small, so that we should state that the work rests on the assumption that the shipping times are sampled from a population which has roughly the shape of a normal distribution.

     1.   <u>Hypotheses</u>

$$H_0 : \mu = 9.5$$
$$H_A : \mu > 9.5$$

2. Level of significance
$$\alpha = 0.01$$

3. Criterion
Reject the null hypothesis if $t \geq 2.998$, where 2.998 is the value of $t_{0.01}$ for $8 - 1 = 7$ degrees of freedom; otherwise, accept H_0 or reserve judgement.

4. Calculations
Solving for s using the 8 items in the sample we get $\bar{x} = 13$ and $s = 3.207$, and then
$$t = \frac{13 - 9.5}{\frac{3.207}{\sqrt{8}}} \approx 3.09$$

5. Decision
Since $t \approx 3.09$ is greater than 2.998, the null hypothesis must be rejected; we conclude that the average time to fill orders exceeds 9.5 days.
Since $t_{.01} = 2.998$ and $t_{.005} = 3.499$, and $t = 3.09$ falls between these two values, we can write $0.005 < p < 0.01$.

10.29 (a) The test statistic is
$$t = \frac{237 - 250}{\frac{25}{\sqrt{20}}} \approx -2.33$$

(b) This is a one-sided test, and t has 19 degrees of freedom. Since
$$-2.539 = -t_{0.01} < -2.33 < -t_{0.025} = -2.093$$
we say that $0.01 < p < 0.025$.

10.31 (a) The probability that an individual medicine will be found ineffective is 0.90. The probability that 20 will all be found ineffective is $0.90^{20} \approx 0.1216$. Hence the probability that at least one will be found effective is $1 - 0.90^{20} \approx 0.8784$.

(b) The probability that none will be found effective is $(0.90)^{20} \approx 0.1216$, the probability that one will be found effective is $20(0.10)(0.90)^{19} \approx 0.2702$, and the probability that more than one will be found effective is approximately $1 - 0.1216 - 0.2702 = 0.6082$.

10.33 Computer exercise.

10.35 Practice exercise.

10.37 1. Hypotheses
$$H_0 : \mu_1 = \mu_2$$
$$H_A : \mu_1 \neq \mu_2$$

2. Level of significance
$$\alpha = 0.01$$

3. Criterion
Reject H_0 if $z \leq -2.575$ or $z \geq 2.575$; otherwise accept the null hypothesis or reserve judgment, where
$$z = \frac{\bar{x}_1 - \bar{x}_2}{\sqrt{\frac{\sigma_1^2}{n_1} + \frac{\sigma_2^2}{n_1}}}$$
with s_1 and s_2 substituted for σ_1 and σ_2.

4. Calculations

$$z = \frac{99 - 96}{\sqrt{\frac{(6.1)^2}{50} + \frac{(6.4)^2}{50}}} = 2.40$$

5. Decision
Since $z = -2.40$ falls between -2.575 and 2.575 the null hypothesis cannot be rejected. Since $p = 0.0164$, which is greater than $\alpha = 0.01$, we cannot reject the null hypothesis.

10.39 (a) $2(0.5000 - 0.4783) = 0.0434$ (e) $2(0.5000 - 0.2088) = 0.5824$
(b) $2(0.5000 - 0.4783) = 0.0434$ (f) $2(0.5000 - 0.4974) = 0.0052$
(c) $2(0.5000 - 0.4761) = 0.0478$ (g) $2(0.5000 - 0.000) = 1.0000$
(d) $2(0.5000 - 0.4761) = 0.0478$ (h) $2(0.50000 - 0.4999997) = 0.0000006$

10.41 The probability that none of the research groups will reject the true null hypothesis that there is no difference between the advertised patent medicine and the competing brand is $(1 - 0.05)^{20} = 0.95^{20}$. The probability that at least one research group will get a spurious result which the advertiser wants is $1 - (0.95)^{20} = 1.0 - 0.36 = 0.64$.

10.43 Practice exercise.

10.45 Practice exercise

10.47 The means of the two samples are 80.0 and 84.0, and their standard deviations are approximately 3.37 and 5.40. Since

$$s_p = \sqrt{\frac{9(3.37)^2 + 9(5.40)^2}{18}} \approx 4.50$$

and since

$$t = \frac{80.0 - 84.0}{4.50\sqrt{\frac{1}{10} + \frac{1}{10}}} \approx -1.99$$

falls between -2.878 and 2.878, the null hypothesis cannot be rejected. The difference between the two sample means is not significant.

10.49 Practice exercise.

10.51 (a) The mean of the five samples are $\bar{x}_1 = \frac{108}{4} = 27$, $\bar{x}_2 = \frac{112}{4} = 28$, $\bar{x}_3 = \frac{136}{4} = 34$,
$\bar{x}_4 = \frac{128}{4} = 32$, and $\bar{x}_5 = \frac{116}{4} = 29$, and their mean is $\frac{(27 + 28 + 34 + 32 + 29)}{5} = \frac{150}{5} = 30$,
so that

$$s_{\bar{x}}^2 = \frac{(27-30)^2 + (28-30)^2 + (34-30)^2 + (32-30)^2 + (29-30)^2}{5-1} = \frac{34}{4} = \frac{17}{2} \text{ and}$$
$$n \cdot s_{\bar{x}}^2 = 4 \cdot \frac{17}{2} = 34.$$

Also,

$$s_1^2 = \frac{(30-27)^2 + (25-27)^2 + (27-27)^2 + (26-27)^2}{4-1} = \frac{14}{3}$$

$$s_2^2 = \frac{(29-28)^2 + (26-28)^2 + (29-28)^2 + (28-28)^2}{4-1} = 2$$

$$s_3^2 = \frac{(32-34)^2 + (32-34)^2 + (35-34)^2 + (37-34)^2}{4-1} = 6$$

$$s_4^2 = \frac{(29-32)^2 + (34-32)^2 + (32-32)^2 + (33-32)^2}{4-1} = \frac{14}{3}$$

$$s_5^2 = \frac{(32-29)^2 + (26-29)^2 + (31-29)^2 + (27-29)^2}{4-1} = \frac{26}{3}$$

so that $\dfrac{s_1^2 + s_2^2 + s_3^2 + s_4^2 + s_5^2}{5} = \dfrac{\frac{14}{3} + 2 + 6 + \frac{14}{3} + \frac{26}{3}}{5} = \dfrac{26}{5} = 5.2$. Then $F = \dfrac{34}{5.2} \approx 6.54$.

(b) Since $F = 6.54$ exceeds 4.89, the value of $F_{0.01}$ for $5 - 1 = 4$ and $5(4 - 1) = 15$ degrees of freedom the null hypothesis must be rejected. The differences among the sample means are significant.

10.53   1.   <u>Hypotheses</u>

$$H_0 : \mu_1 = \mu_2 = \mu_3$$
$$H_A : \text{The } \mu\text{'s are not all equal}$$

2.   <u>Level of significance</u>

$$\alpha = 0.01$$

3.   <u>Criterion</u>

Reject the null hypothesis if $F \geq 5.78$, the value of $F_{0.01}$ for $3 - 1 = 2$ and $3(8 - 1) = 21$ degrees of freedom; otherwise accept H_0 or reserve judgment.

4.   <u>Calculations</u>

Since $T_1 = 2{,}368$, $T_2 = 2{,}346$, $T_3 = 2{,}353$, $T.. = 7{,}067$, and $\sum\sum x_{ij}^2 = 2{,}085{,}151$ we get

$$SST = 2{,}085{,}151 - \frac{(7{,}067)^2}{24} \approx 2{,}085{,}151 - 2{,}080{,}937.0417 = 4{,}213.9583,$$

$$SS(Tr) = \frac{2{,}368^2 + 2{,}346^2 + 2{,}353^2}{8} - 2{,}080{,}937.0417,$$

$$= 2{,}080{,}968.6250 - 2{,}080{,}937.0417 = 31.5833$$

and $SSE = 4{,}213.9583 - 31.5833 = 4{,}182.3750$. The remainder of the arithmetic can be displayed in an analysis of variance table:

Source of variation	Degrees of freedom	Sum of squares	Mean square	F
Treatments	2	31.5833	15.7916	0.08
Error	21	4,182.3750	199.1607	
Total	23	4,213.9583		

5.   <u>Decision</u>

Since $F = 0.08$ does not exceed 5.78, the null hypothesis cannot be rejected. The differences among the average weekly earnings may be attributed to chance. The p-value exceeds 0.05; we can write $p > 0.05$.

10.55 1. Hypotheses

$$H_0 : \mu_1 = \mu_2 = \mu_3 = \mu_4$$
$$H_A : \text{The } \mu\text{'s are not all equal}$$

2. Level of significance

$$\alpha = 0.01$$

3. Criterion

Reject the null hypothesis if $F \geq 4.68$. We seek the value of $F_{0.01}$ for $4 - 1 = 3$ and $N - k = 31 - 4 = 27$ degrees of freedom; since this is not available in the table we use the conservatively chosen larger value for 3 and 25 degrees of freedom.

4. Calculations

Since $T_1 = 574$, $T_2 = 547$, $T_3 = 449$, $T_4 = 584$, $T.. = 2,154$, and $\sum \sum x_{ij}^2 = 150,624$ we

get $SST = 150,624 - \dfrac{(2,154)^2}{31} \approx 150,624 - 149,668.2581 = 955.7419$.

$SS(Tr) = \dfrac{574^2}{8} + \dfrac{547^2}{8} + \dfrac{449^2}{6} + \dfrac{584^2}{9} - 149,668.2581 \approx 150,080.9028 - 149,668.2581 = 412.6447$, and $SSE = 955.7419 - 412.6447 = 543.0972$. The remainder of the arithmetic can be displayed in an analysis of variance table:

Source of variation	Degrees of freedom	Sum of squares	Mean square	F
Treatments	3	412.6447	137.5482	6.84
Error	27	543.0972	20.1147	
# Total	30	955.7419		

5. Decision

Since $F = 6.84$ exceeds 4.68, the null hypothesis must be rejected; we conclude that the secretary does not type with equal speed on the four typewriters.
Here we note $p < 0.01$.

10.57 Computer exercise.

10.59 Computer exercise.

Chapter 11

Tests Based on Count Data

11.1   Practice exercise.

11.3   1. $H_0 : p = 0.60$
             $H_A : p < 0.60$
         2. $\alpha = 0.05$
         3.   The test statistic is x, which is 5.
         4.   Table I shows that for $n = 15$ and $p = 0.60$ the probability of five or fewer successes is $0.002 + 0.007 + 0.024 = 0.033$.
         5.   Since 0.033 is less than 0.05 the null hypothesis must be rejected. The data refutes the claim.

11.5   1. $H_0 : p = 0.40$
             $H_A : p \neq 0.40$
         2. $\alpha = 0.01$
         3.   The test statistic is x, which is 3.
         4.   Table I shows that for $n = 10$ and $p = 0.40$ the probability of three or fewer successes is $0.006 + 0.040 + 0.121 + 0.215 = 0.382$, and the probability of three or more successes is $0.215 + 0.251 + 0.201 + 0.111 + 0.042 + 0.011 + 0.002 = 0.833$.
         5.   Since neither probability is less than or equal to 0.05, the null hypothesis cannot be rejected.

11.7   1. $H_0 : p = 0.90$
             $H_A : p < 0.90$
         2. $\alpha = 0.05$
         3.   The test statistic is x, which is 12.
         4.   Table I shows that for $n = 15$ and $p = 0.90$ the probability of twelve or fewer successes is $0.002 + 0.010 + 0.043 + 0.129 = 0.184$.
         5.   Since 0.184 is not less than or equal to 0.05, the null hypothesis cannot be rejected.

11.9   1. $H_0 : p = 0.50$
             $H_A : p < 0.50$
         2. $\alpha = 0.01$
         3.   Reject the null hypothesis if $z < -2.33$; otherwise accept the null hypothesis or reserve judgment.
         4. $z = \dfrac{85 - 200(0.50)}{\sqrt{200(0.50)(0.50)}} \approx -2.12$
         5.   Since -2.12 is greater than -2.33, the null hypothesis cannot be rejected. The data does not refute the claim.

11.11 1. $H_0 : p = 0.30$
$H_A : p > 0.30$
2. $\alpha = 0.01$
3. Reject the null hypothesis if $z \geq 2.33$; otherwise accept H_0 or reserve judgment.
4. $z = \dfrac{169 - 500(0.30)}{\sqrt{500(0.30)(0.70)}} \approx 1.85$
5. Since 1.85 does not exceed 2.33, the null hypothesis cannot be rejected.

11.13 1. $H_0 : p = 0.15$
$H_A : p < 0.15$
2. $\alpha = 0.05$
3. Reject the null hypothesis if $z \leq -1.645$; otherwise accept H_0 or reserve judgment.
4. $z = \dfrac{32 - (300)(0.15)}{\sqrt{300(0.15)(1 - 0.15)}} = \dfrac{-13}{6.185} \approx -2.10$
5. Since -2.10 is less than -1.645, the null hypothesis must be rejected. We conclude that fewer than 0.15 of the shoppers make no purchase, or we reserve judgment.

11.15 1. Hypotheses
$H_0 : p_1 = p_2$
$H_A : p_1 > p_2$
2. Level of significance
$\alpha = 0.01$
3. Criterion
Reject the null hypothesis if $z \geq 2.33$ using the formula identified as "Statistic for a large sample test concerning the difference between two proportions."
4. Calculations
Substituting $x_1 = 56$, $x_2 = 38$, $n_1 = 80$, $n_2 = 80$, and $\hat{p} = \dfrac{56 + 38}{80 + 80} = 0.5875$ into the

formula for z, we get $z = \dfrac{\dfrac{56}{80} - \dfrac{38}{80}}{\sqrt{(0.5875)(0.4125)\left(\dfrac{1}{80} + \dfrac{1}{80}\right)}} \approx 2.89$

5. Decision
Since $z = 2.89$ exceeds 2.33, the null hypothesis must be rejected; in other words, we conclude that the new pain-relieving drug is effective.

11.17 1. $H_0 : p_1 = p_2$
$H_A : p_1 \neq p_2$
2. $\alpha = 0.05$
3. Reject the null hypothesis if $z \leq -1.96$ or $z \geq 1.96$ using the formula identified as "Statistic for a large sample test concerning the difference between two proportions." Otherwise accept the null hypothesis or reserve judgment.
4. Substituting $x_1 = 28$, $n_1 = 400$, $x_2 = 15$, $n_2 = 300$, and $\hat{p} = \dfrac{28 + 15}{400 + 300} = 0.061$, we get

$$z = \frac{\dfrac{28}{400} - \dfrac{15}{300}}{\sqrt{(0.061)(0.939)\left(\dfrac{1}{400} + \dfrac{1}{300}\right)}} = 1.09$$

5.   Since this value falls between $-z_{0.025} = -1.96$ and $z_{0.025} = 1.96$, we find that the null hypothesis cannot be rejected. In other words we cannot conclude that there is a real difference between the true proportions of defectives. We, therefore, accept the null hypothesis or reserve judgment.

11.19   Practice exercise.

11.21   1. $H_0 : p_1 = p_2 = p_3$
              $H_A :$ The p's are not all equal.

        2. $\alpha = 0.05$

        3.   Reject the null hypothesis if $\chi^2 = 5.991$, which is the value of χ^2 for $3 - 1 = 2$ degrees of freedom; otherwise, accept H_0 or reserve judgment.

        4.   Since $\dfrac{32 + 69 + 19}{200 + 300 + 100} = \dfrac{120}{600} = 0.20$, the expected frequencies for the first row are $200(0.20) = 40$, $300(0.20) = 60$, and $100(0.20) = 20$, and the expected frequencies for the second row are $200 - 40 = 160$, $300 - 60 = 240$, and $100 - 20 = 80$. Therefore,

$$\chi^2 = \frac{(32-40)^2}{40} + \frac{(69-60)^2}{60} + \frac{(19-20)^2}{20} + \frac{(168-160)^2}{160} + \frac{(231-240)^2}{240} + \frac{(81-80)^2}{80}$$
$$= 3.75$$

        5.   Since $\chi^2 = 3.75$ is less than 5.991, the null hypothesis cannot be rejected. Accept the null hypothesis or reserve judgment.

11.23   1. $H_0 : p_1 = p_2 = p_3 = p_4 = p_5$
              $H_A :$ The p's are not all equal.

        2. $\alpha = 0.01$

        3.   Reject the null hypothesis if $\chi^2 \geq 13.277$, which is the value of $\chi^2_{0.01}$ for $5 - 1 = 4$ degrees of freedom; otherwise, accept H_0 or reserve judgment.

        4.   Since $\hat{p} = \dfrac{74 + 81 + 69 + 75 + 91}{100 + 100 + 100 + 100 + 100} = 0.78$, the expected frequencies for the first row are $100(0.78) = 78$, and the expected frequencies for the second row are $100 - 78 = 22$. Therefore,

$$\chi^2 = \frac{(74-78)^2}{78} + \frac{(81-78)^2}{78} + \frac{(69-78)^2}{78} + \frac{(75-78)^2}{78}$$
$$+ \frac{(91-78)^2}{78} + \frac{(26-22)^2}{22} + \frac{(19-22)^2}{22} + \frac{(31-22)^2}{22} + \frac{(25-22)^2}{22} + \frac{(9-22)^2}{22} = 16.55.$$

        5.   Since $\chi^2 = 16.55$ exceeds 13.277, the null hypothesis must be rejected; we conclude that the true proportions of members of the five unions who are for the legislation are not all equal.

11.25   1.   Hypothesis

$H_0 : p_1 = p_2$

$H_A : p_1 \neq p_2$

2.   Level of significance

$\alpha = 0.05$

3.   Criterion

Reject the null hypothesis if $\chi^2 \geq 3.841$, the value of $\chi^2_{0.05}$ for $2 - 1 = 1$ degree of freedom, where

$$\chi^2 = \sum \frac{(o - e)^2}{e}$$

otherwise, accept it or reserve judgment.

4.   Calculations

Since $\hat{p} = \dfrac{141 + 123}{600} = 0.44$, we get $300(0.44) = 132$ for both expected frequencies for the first row, and $300 - 132 = 168$ for both expected frequencies for the second row. Then, substituting into the formula for χ^2, we get

$$\chi^2 = \frac{(141 - 132)^2}{132} + \frac{(123 - 132)^2}{132} + \frac{(159 - 168)^2}{168} + \frac{(177 - 168)^2}{168} \approx 2.19$$

5.   Decision

Since $\chi^2 = 2.19$ does not exceed 3.841, the null hypothesis cannot be rejected; in other words, the difference between the two sample proportions is not significant.

If we square 1.48, the value obtained for z in Exercise 11.16, we get $(1.48)^2 = 2.1904$, and except for rounding, this equals the value which we obtained for χ^2.

11.27   Practice exercise.

11.29   Practice exercise.

11.31   Practice exercise.

11.33   Computer exercise.

11.35   Computer exercise.

11.37   Computer exercise.

11.39   Practice exercise.

11.41   Practice exercise.

11.43   Practice exercise.

11.45   Practice exercise.

R.119 The differences are $-2, 4, 0, 1, 2, -1, 2, -1, 1, 0, 3, -1, 0, -2, 3, 2, 0$, and 4, and for these data we perform the following test:

1. $H_0 : \mu = 0$
    $H_A : \mu \neq 0$
2. $\alpha = 0.05$
3. Reject the null hypothesis if $t \leq -2.110$ or $t \geq 2.110$, where 2.110 is the value of $t_{0.025}$ for $18 - 1 = 17$ degrees of freedom; otherwise, accept H_0 or reserve judgment.
4. $n = 18$, $\sum x = 15$, and $\sum x^2 = 75$, so that $\bar{x} = \frac{15}{18} \approx 0.833$ and $S_{xx} = 75 - \frac{15^2}{18} = 62.5$.

    Then $s = \sqrt{\frac{62.5}{17}} \approx 1.917$ and $t = \dfrac{0.833}{\frac{1.917}{\sqrt{18}}} \approx 1.84$.
5. Since $t = 1.84$ falls between -2.110 and 2.110, the null hypothesis cannot be rejected; the difference is not significant.

R.121 The sample proportion is $\dfrac{258}{300} = 0.86$, and its standard error is $\sqrt{\dfrac{(0.86)(0.14)}{300}} \approx 0.0200$

The 95% confidence interval is $0.86 \pm 1.96 \cdot 0.02$ or $0.086 \pm 0.0392 \approx (0.8208, 0.8992)$. This interval may be written as $(0.82, 0.90)$.

R.123 (a) 1.812;
     (b) 2.201;
     (c) 2.681;
     (d) 3.012.

R.125 Substituting $n = 64$, $\bar{x} = 2.60$, $s = 0.40$, and $z_{0.025} = 1.96$, into the formula for a large sample confidence interval for μ.

$$\bar{x} - z_{\frac{\alpha}{2}} \cdot \frac{\sigma}{\sqrt{n}} < \mu < \bar{x} + z_{\frac{\alpha}{2}} \cdot \frac{\sigma}{\sqrt{n}}, \text{ we get } 2.60 - 1.96 \cdot \frac{0.40}{\sqrt{64}} < \mu < 2.60 + 1.96 \cdot \frac{0.40}{\sqrt{64}} \text{ and}$$

$2.50 < \mu < 2.70$.

R.127 Substituting $s = 0.40$, $n = 64$, and $z_{\frac{\alpha}{2}} = 1.96$ into the formula for a large sample confidence

interval for σ, $\dfrac{s}{1 + \frac{z_{\frac{\alpha}{2}}}{\sqrt{2n}}} < \sigma < \dfrac{s}{1 - \frac{z_{\frac{\alpha}{2}}}{\sqrt{2n}}}$, we get $\dfrac{0.40}{1 + \frac{1.96}{\sqrt{2(64)}}} < \sigma < \dfrac{0.40}{1 - \frac{1.96}{\sqrt{2(64)}}}$ and

$0.34 < \sigma < 0.48$.

R.129 The maximum error is $E = 2.575 \cdot \dfrac{2.1}{\sqrt{100}} \approx 0.541$.

R.131 (a) $126.5 - 1.96 \cdot \dfrac{26.4}{\sqrt{100}} < \mu < 126.5 + 1.96 \cdot \dfrac{26.4}{\sqrt{100}}$ so that 121.33 minutes $< \mu < 131.67$ minutes.

(b) $126.5 - 2.575 \cdot \dfrac{26.4}{\sqrt{100}} < \mu < 126.5 + 2.575 \cdot \dfrac{26.4}{\sqrt{100}}$ so that 119.70 minutes $< \mu < 133.30$ minutes.

R.133 $125.45 - 1.96 \cdot \dfrac{37.15}{\sqrt{40}} < \mu < 126.45 + 1.96 \cdot \dfrac{37.15}{\sqrt{40}}$ so that $\$114.94 < \mu < \$137.96$.

R.135 $\dfrac{1.29}{1 + \dfrac{2.33}{\sqrt{200}}} < \sigma < \dfrac{1.29}{1 - \dfrac{2.33}{\sqrt{200}}}$, $\dfrac{1.29}{1.16476} < \sigma < \dfrac{1.29}{0.83524}$ and 1.11 minutes $< \sigma < 1.54$ minutes.

R.137 Since $\dfrac{x}{n} = \dfrac{28}{80} = 0.35$ and $z_{0.005} = 2.575$, the maximum error is $E = 2.575\sqrt{\dfrac{(0.35)(0.65)}{80}} \approx$ 0.137.

R.139 Since $\dfrac{x}{n} = \dfrac{979}{1,419} \approx 0.690$ and $z_{0.005} = 2.575$, we get

$0.690 - 2.575\sqrt{\dfrac{(0.690)(0.310)}{1,419}} < p < 0.690 + 2.575\sqrt{\dfrac{(0.690)(0.310)}{1,419}}$, and hence $0.658 < p < 0.722$.

R.141 (a) She is testing the null hypothesis that the average value is not $195,000.
(b) She is testing the null hypothesis that the average value is $195,000.

R.143 1. <u>Hypotheses</u>
$$H_0: \ \mu = 81.7$$
$$H_A: \ \mu \neq 81.7$$
This investigation seems to have been conducted for the simple purpose of comparing these students to the eighth grade norm. The fact that the students scored below the norm should not cause you to make this into a one-tail test. In other words, the numerical values of the data should not dictate the formulation of the hypotheses.

2. <u>Level of significance</u>
$$\alpha = 0.05$$

3. <u>Criterion</u>
Reject H_0 is $z \leq -1.96$ or $z \geq +1.96$.

4. <u>Calculations</u>
$$z = \frac{79.6 - 81.7}{\dfrac{8.5}{\sqrt{100}}} \approx -2.47$$

5. Decision

Since $z = -2.47 < -1.96$, we reject the null hypothesis.

Since $z = -2.47$ and since the test is two-sided, we find that the p-value is $2(0.0068) = 0.0136$. This is well below 5%, and it indicates that we should reject the null hypothesis.

R.145 1. Hypotheses

$$H_0: \ \mu_1 = \mu_2$$
$$H_A: \ \mu_1 > \mu_2$$

2. Level of significance

$$\alpha = 0.01$$

3. Criterion

Reject the null hypothesis if $z \geq 2.33$, where

$$z = \frac{\bar{x}_1 - \bar{x}_2}{\sqrt{\dfrac{\sigma_1^2}{n_1} + \dfrac{\sigma_2^2}{n_1}}}$$

with s_1 and s_2 substituted for σ_1 and σ_2. Otherwise, state that the difference between the two sample means is not significant.

4. Calculations

$$z = \frac{8.6 - 8.3}{\sqrt{\dfrac{(0.75)^2}{70} + \dfrac{(0.80)^2}{50}}} \approx 2.08$$

5. Decision

Since 2.08 is less than 2.33, the null hypothesis cannot be rejected. We accept H_0 or reserve judgment.

The p-value for 2.08 (which the exercise does not ask for) is $0.5000 - 0.4812 = 0.0188$. Since this is greater than 0.01, the conclusion is the same as before.

R.147 The differences are 11, 12, 7, 23, 18, 18, 8, 3, 17, 16, 8, and 18. Their mean is 13.25 and their standard deviation is approximately 5.956.

Since

$$t = \frac{13.25 - 10}{\dfrac{5.956}{\sqrt{12}}} = 1.89$$

is greater than 1.796, the null hypothesis must be rejected. The medication reduces the blood sugar by more than 10 points in six weeks.

R.149 1. $H_0: \ p = 0.80$

$H_A: \ p < 0.80$

2. $\alpha = 0.05$

3.' The test statistic is x, which is 9.

4.' Table I shows that for $n = 15$ and $p = 0.80$ the probability of nine or fewer successes is $0.001 + 0.003 + 0.014 + 0.043 = 0.061$.

5.' Since 0.061 is greater than 0.05, the null hypothesis cannot be rejected. The data do not refute the claim.

R.151 1. H_0: $p = 0.60$
            H_A: $p \neq 0.60$
         2. $\alpha = 0.05$
         3.' The test statistic is x, which is 6.
         4.' Table I shows that for $n = 15$ and $p = 0.60$ the probability of six or fewer successes is $0.002 + 0.007 + 0.024 + 0.061 = 0.094$, and that the probability of six or more successes is $1 - (0.002 + 0.007 + 0.024) = 0.967$
         5.' Since neither probability is less than or equal to 0.025, the null hypothesis cannot be rejected.

R.153 (a) 1. H_0: $p = 0.90$
                H_A: $p \neq 0.90$
             2. $\alpha = 0.05$
             3.' The test statistic is x, which is 11.
             4.' Table I shows that for $n = 15$ and $p = 0.90$ the probability of eleven or fewer successes is 0.055, and the probability of eleven or more successes is 0.988.
             5.' Since neither probability is less than or equal to 0.025, the null hypothesis cannot be rejected.
        (b) This is similar to (a), but with $x = 15$, the probability of fifteen or fewer successes is 1.000, while the probability of fifteen or more successes is 0.206–and the null hypothesis cannot be rejected. Thus, getting a 100% success rate in the experiment does not lead us to conclude that $p \neq 0.90$.

R.155 1. H_0: $p_1 = p_2$
            H_A: $p_1 > p_2$
         2. $\alpha = 0.05$
         3.' Reject the null hypothesis if $z \geq 1.645$, otherwise accept H_0 or reserve judgment.
         4.' Since $\hat{p} = \dfrac{246 + 165}{300 + 300} = 0.685$, $z = \dfrac{\dfrac{246}{300} - \dfrac{165}{300}}{\sqrt{(0.685)(0.315)\left(\dfrac{1}{300} + \dfrac{1}{300}\right)}} \approx \dfrac{0.270}{0.03793} \approx 7.119$
         5.' Since $z = 7.119$ exceeds 1.645, the null hypothesis must be rejected; we conclude that the test item discriminates between well-prepared and poorly prepared students.

Chapter 12

Regression and Correlation

12.1 Practice exercise.

12.3

Number of Employees	Number of Cups of Coffee		
x	y	x^2	xy
11	18	121	198
13	36	169	468
15	40	225	600
18	50	324	900
21	58	441	1,218
24	74	576	1,776
102	276	1,856	5,160

(a) $n = 6$, $\sum x = 102$, $\sum y = 276$, $\sum xy = 5{,}160$, $\bar{x} = \dfrac{102}{6} = 17$, and $\bar{y} = 46$

(b) $S_{xx} = 1{,}856 - \dfrac{(102)^2}{6} = 122$ and $S_{xy} = 5{,}160 - \dfrac{(102)\cdot(276)}{6} = 468$

(c) $b = \dfrac{468}{122} = 3.836$, and $a = 46 - (3.836)\cdot(17) = -19.212$

(d) $a + bx = -19.212 + 3.836 \cdot x$

(e) $-19.212 + 76.720 = 57.508$ cups of coffee.

12.5

Age	Height		
x	y	x^2	xy
13	62	169	806
11	56	121	616
11	58	121	638
6	46	36	276
9	51	81	459
7	50	49	350
10	54	100	540
67	377	677	3,685

Find b first, since a is calculated in terms of b.

$S_{xx} = 677 - \dfrac{(67)^2}{7} \approx 35.714$; $\quad S_{xy} = 3{,}685 - \dfrac{(67)(377)}{7} \approx 76.571$; and $b = \dfrac{76.571}{35.714} \approx 2.144$ and

$a = 53.857 - (2.144)(9.571) \approx 33.337$.

The equation is $33.337 + 2.144(x)$, and for 12 years of age we get $\hat{y} = 33.337 + 2.144(12) \approx 59.07$ inches.

12.7 Numbering the years as 1, 2, 3, 4, and 5, we get $n = 5$, $\sum x = 15$, $\sum x^2 = 55$, $\sum y = 12.3$,

and $\sum xy = 42.9$, so that $\bar{x} = 3$, $\bar{y} = 2.46$, $S_{xx} = 55 - \dfrac{15^2}{5} = 10$, and $S_{xy} = 42.9 - \dfrac{(15)(12.3)}{5}$

$= 6$. Then $b = \dfrac{6}{10} = 0.6$ and $a = 2.46 - (0.6)(3) = 0.66$ and the equation is $\hat{y} = 0.66 + 0.6x$.

Therefore, the company's predicted income from sales during the sixth year of operation is $\hat{y} = 0.66 + 0.6(6) = 4.26$, or $4,260,000.

12.9 $\hat{y} = 70.950 + 1.126x$.

12.11 Practice exercise.

12.13 1. $H_0 : \beta = 0.10$

          $H_A : \beta < 0.10$

     2. $\alpha = 0.01$

     3.   Reject the null hypothesis if $t \le -3.747$ where 3.747 is the value of $t_{0.01}$ for $6 - 2 = 4$ degrees of freedom; otherwise, accept H_0 or reserve judgment.

     4.   Since $n = 6$, $\sum x = 90$, $\sum x^2 = 1{,}694$, $\sum y = 10.5$, $\sum y^2 = 20.29$, and $\sum xy = 181.1$,

         we get $\bar{x} = 15$, $\bar{y} = 1.75$, $S_{xx} = 1{,}694 - \dfrac{90^2}{6} = 344$, $S_{yy} = 20.29 - \dfrac{10.5^2}{6} = 1.915$, and

         $S_{xy} = 181.1 - \dfrac{(90)(10.5)}{6} = 23.6$. Then $b = \dfrac{23.6}{344} \approx 0.0686$ and $a = 1.75 - 0.0686(15) =$

         0.7210. Also $s_e = \sqrt{\dfrac{1.915 - \dfrac{23.6^2}{344}}{6 - 2}} \approx 0.2720$. Then $t = \dfrac{0.0686 - 0.10}{0.2720} \sqrt{344} \approx -2.14$.

     5.   Since $t = -2.14$ is not less than -3.747, the null hypothesis cannot be rejected; the difference between $b = 0.686$ and the target value 0.10 is not significant.

12.15 Use the values from problem 12.3, along with $\sum y^2 = 14{,}560$ and

$S_{yy} = 14{,}560 - \dfrac{(276)^2}{6} = 1{,}864$. Then find $s_e = \sqrt{\dfrac{1{,}864 - \dfrac{(468)^2}{122}}{6 - 2}} \approx 4.145$. The confidence

interval is $3.836 \pm 4.604\left(\dfrac{4.145}{\sqrt{122}}\right)$, which is 3.836 ± 1.7277; hence $2.108 < \beta < 5.564$.

12.17   The prediction is $6.0281 - 1.4927(60) \approx 95.590$. Using the quantities computed in Exercise

12.12, the 95% confidence interval is $95.590 \pm (2.447)(2.7560)\sqrt{\frac{1}{8} + \frac{(60 - 50.9375)^2}{2,383.48}}$ which

is 95.590 ± 2.693. The interval may be given as $(92.897, 98.283)$.

12.19 $t = 0.250$ and the null hypothesis cannot be rejected.

12.21 $0.177 < \beta < 2.076$.

12.23   Practice exercise.

12.25   These data have $n = 9$, $\sum x = 544$, $\sum x^2 = 33,694$, $\sum y = 4,517$, $\sum y^2 = 2,347,463$, and

$\sum xy = 275,872$. Then $S_{xx} = 33,694 - \frac{544^2}{9} \approx 812.22$, $S_{yy} = 2,347,463 - \frac{4,517^2}{9} \approx$

80,430.89, and $S_{xy} = 275,872 - \frac{(544)(4,517)}{9} \approx 2,844.44$. We then get

$r = \dfrac{2,844.44}{\sqrt{(812.22)(80,430.89)}} \approx 0.352$.

12.27   For this problem find $n = 7$, $\sum x = 324$, $\sum x^2 = 15,516$, $\sum y = 327$, $\sum y^2 = 15,967$, and

$\sum xy = 15,666$ Then $S_{xx} = 15,516 - \frac{324^2}{7} \approx 519.429$, $S_{yy} = 15,967 - \frac{327^2}{7} \approx 691.429$, and

$S_{xy} = 15,666 - \frac{(324)(327)}{7} \approx 530.571$. We then get $r = \dfrac{530.571}{\sqrt{(519.429)(691.429)}} \approx 0.885$.

12.29   For this problem find $n = 16$, $\sum x = 1,127$, $\sum x^2 = 85,589$, $\sum y = 1,244$, $\sum y^2 = 100,326$,

and $\sum xy = 91,959$. Then $S_{xx} = 85,589 - \frac{1,127^2}{16} \approx 6,205.938$,

$S_{yy} = 100,326 - \frac{1,244^2}{16} = 3,605$, and $S_{xy} = 91,959 - \frac{(1,127)(1,244)}{16} = 4,334.750$. We then get

$r = \dfrac{4,334.750}{\sqrt{(6,205.938)(3,605)}} \approx 0.916$.

12.31   Practice exercise.

12.33   In Exercise 12.14, we found $S_{xx} \approx 2.2222$, $S_{yy} = 1,081,088.89$, and $S_{xy} = -1,039.4444$. We

then get $r = \dfrac{-1,039.4444}{\sqrt{(2.2222)(1,081,088.89)}} \approx -0.6706$. It follows that $(-0.6706)^2(100\%)$

$= 45.0\%$. of the variation of the weekly profit can be attributed to differences in price.

12.35 It should not come as a surprise since we can draw a straight line through any two points and, hence, get a perfect fit.

    (a)   We get $r = 1$ since the larger value of x corresponds to the larger value of y.

    (b)   We get $r = -1$ since the larger value of x corresponds to the smaller value of y.

12.37 (a)   Since $r = 0.62$ exceeds 0.590, where 0.590 is the value of $r_{0.005}$ for $n = 18$, it is significant.

    (b)   Since $r = -0.47$ is less than 0.449, where 0.449 is the value of $r_{0.005}$ for $n = 32$, it is significant.

    (c)   Since $r = -0.58$ falls between -0.623 and 0.623 where 0.623 is the value of $r_{0.005}$ for $n = 16$, it is not significant.

    (d)   Since $r = 0.63$ falls between -0.661 and 0.661, where 0.661 is the value of $r_{0.005}$ for $n = 14$, it is not significant.

12.39 $SST = 2,660.64$, $SSR = 1,957.192$, and $r = 0.514$.

12.41 $r \approx 0.916$, so that $(0.916)^2(100\%) \approx 83.9\%$ of the variation of the anthropology scores can be attributed to the relationship between students' grades in the two subjects.

Chapter 13

13.1   Practice exercise.

13.3   The value which equals 5 has to be discarded, so that the sample size is only $n = 11$.
1. H_0: $\tilde{\mu} = 5$
     H_A: $\tilde{\mu} > 5$
2. $\alpha = 0.05$
3.   The statistic is x, the number of plus signs.
4.   Replacing each value greater than 5 with a plus sign and each value less than 5 with a minus sign, we get, $+ + - + + - - + + + +$ so that $x = 8$.
     From Table I with $n = 11$, and $p = 0.50$, the probability of 8 or more successes is $0.081 + 0.027 + 0.005 = 0.113$.
5.   Since 0.113 exceeds 0.050, the null hypothesis cannot be rejected.

13.5   Since the difference is zero for one of the pairs, the sample size is reduced to 14.
1. H_0: $\mu_1 = \mu_2$
     H_A: $\mu_1 > \mu_2$
2. $\alpha = 0.01$
3.   If x is the number of plus signs (the first fire department responds to more alarms than the second fire department), reject the null hypothesis if the probability of getting x or more plus signs is less than or equal to 0.01; otherwise accept it or reserve judgment.
4.   Replacing each pair of values with a plus sign if the first value is greater than the second, we get 9 positive values (+). Thus, $x = 9$ and Table I shows that for $n = 14$ and $p = 0.5$ the probability of 9 or more plus signs is $0.122 + 0.061 + 0.022 + 0.006 + 0.001 = 0.212$
5.   Since 0.212 exceeds 0.01, the null hypothesis cannot be rejected. There is no evidence that there is difference in the number of alarms to which the two fire departments respond at the 0.01 level of significance.

13.7   Practice exercise.

13.9   The two values which equal 100.0 have to be discarded, so that the sample size is only $n = 12$.
1. H_0: $\tilde{\mu} = 100.0$
     H_A: $\tilde{\mu} \neq 100.0$
2. $\alpha = 0.05$
3.   The statistic is x, the number of plus signs.
4.   Replacing each value greater than 100.0 with a plus sign and each value less than 100.0 with a minus sign, we get $+ - + + - + + + - + + +$, so that $x = 9$.
     From Table I with $n = 12$ and $p = 0.50$, the probability of 9 or more successes is $0.054 + 0.016 + 0.003 = 0.073$.
5.   Since 0.073 exceeds 0.05, the null hypothesis cannot be rejected; the data do not refute the claim that the median weight is 100.0 grams per package.

13.11 Since four of the pairs, 0 and 0, 0 and 0, 1 and 1, 1 and 1, must be discarded, the sample size is only $n = 26$.

    1. H_0: $\mu_1 = \mu_2$

         H_A: $\mu_1 > \mu_2$

    2. $\alpha = 0.01$

    3.   Reject the null hypothesis if $z \geq 2.33$; otherwise, accept H_0 or reserve judgment.

    4.   Replacing each positive difference with a plus sign and each negative difference with a minus sign, we get $+ + + - + + + - + + + + + + + - + + - + - + + - + + +$, so that $x = 20$.

$$\text{Therefore, } z = \frac{20 - 26(0.50)}{\sqrt{26(0.50(0.50)}} \approx 2.75.$$

    5.   Since $z = 2.75$ exceeds 2.33, the null hypothesis must be rejected; we conclude that the first archaeologist is better than the second in finding artifacts.

13.13 Since one of the pairs, 6 and 6, must be discarded, the sample size is only $n = 24$.

    1. H_0: $\mu_1 = \mu_2$

         H_A: $\mu_1 < \mu_2$

    2. $\alpha = 0.05$

    3.   Reject the null hypothesis if $z \leq -1.645$; otherwise, accept H_0 or reserve judgment.

    4.   Replacing each positive difference with a plus sign and each negative difference with a minus sign, we get $+ - - - - + - - + - + - - - - + - - + - - + - -$, so that $x = 7$. Therefore,

$$z = \frac{7 - 24(0.50)}{\sqrt{24(0.50)(0.50)}} \approx -2.04$$

    5.   Since $z = -2.04$ is less than -1.645, the null hypothesis must be rejected; we conclude that on the average more employees are absent from the second department.

13.15 1. H_0: The populations are identical.

         H_A: $\mu_1 \neq \mu_2$

    2. $\alpha = 0.05$

    3.   Reject the null hypothesis if $U \leq 8$; which is the value of $U'_{0.05}$ for $n_1 = 6$ and $n_2 = 8$; otherwise, accept H_0 or reserve judgment.

    4.   Arranging the data jointly according to size, we get 56, 58, 63, 63, 70, 72, 74, 75, 76, 77, 80, 82, 85, and 86. Then, if we assign them in this order the ranks 1, 2, ..., and 14, we find that the values of the first sample occupy ranks 1, 2, 3.5, 5, 7, and 8, while those of the second sample occupy ranks 3.5, 6, 9, 10, 11, 12, 13, and 14. Thus, $W_1 = 1 + 2 + 3.5 + 5 + 7 + 8 = 26.5$, $W_2 = 3.5 + 6 + 9 + 10 + 11 + 12 + 13 + 14 = 78.5$, $U_1 = 26.5 - \dfrac{6 \cdot 7}{2} = 5.5$, $U_2 = 78.5 - \dfrac{8 \cdot 9}{2} = 42.5$ and $U = 5.5$.

    5.   Since $U = 5.5$ is less than 8, the null hypothesis must be rejected; we conclude that the average hardness of die castings from the two production lots is not the same.

13.17 1. H_0: The populations are identical.

         H_A: $\mu_1 \neq \mu_2$

    2. $\alpha = 0.05$

    3.   Reject the null hypothesis if $U \leq 5$; which is the value of $U'_{0.05}$ for $n_1 = 6$ and $n_2 = 6$; otherwise, accept H_0 or reserve judgment.

4. Arranging the data according to size, we get 3, 5, 6, 6, 7, 7, 8, 10, 11, 11, 12, and 15. Ranked in this order from 1 through 12, the values of the first sample occupy the ranks 3.5, 5.5, 9.5, 9.5, 11, 12, and those of the second sample occupy ranks 1, 2, 3.5, 5.5, 7, and 8. Thus, $W_1 = 3.5 + 5.5 + 9.5 + 9.5 + 11 + 12 = 51$, $W_2 = 1 + 2 + 3.5 + 5.5 + 7 + 8 = 27$, $U_1 = 51 - \dfrac{6 \cdot 7}{2} = 30$, $U_2 = 27 - \dfrac{6 \cdot 7}{2} = 6$ and $U = 6$.

5. Since $U = 6$ is greater than 5, the null hypothesis cannot be rejected; the difference between the means is not significant.

13.19 Practice exercise.

13.21 1. H_0: The populations are identical.
       H_A: $\mu_1 > \mu_2$
2. $\alpha = 0.01$
3. Reject the null hypothesis if $z \geq 2.33$; otherwise, accept H_0 or reserve judgment.
4. Arranging the data according to size, we get 8.0, 10.9, 11.8, 12.3, 13.5, 13.6, 13.7, 14.3, 14.4, 14.4, 14.6, 14.6, 15.2, 15.2, 17.2, 17.5, 18.1, 18.8, 19.1, 19.3, 20.3, 21.1, 21.2, and 22.5. Ranked in this order from 1 through 24, the values of the first sample occupy ranks 9.5, 11.5, 13.5, 15, 16, 17, 18, 19, 20, 21, 22, and 23, so that $W_1 = 9.5 + 11.5 + 13.5 + 15 + 16 + 17 + 18 + 19 + 20 + 21 = 22 + 23 = 205.5$ and $U_1 = 205.5 - \dfrac{12 \cdot 13}{2} = 127.5$. Since $\mu_{U_1} = \dfrac{12 \cdot 12}{2} = 72$ and $\sigma_{U_1} = \sqrt{\dfrac{12 \cdot 12 \cdot 25}{12}} \approx 17.32$, it follows that
$$z = \frac{127.5 - 72}{17.32} \approx 3.20.$$
5. Since $z = 3.20$ exceeds 2.33, the null hypothesis must be rejected; we conclude that on the average brand X flares last longer than brand Y flares.

13.23 1. H_0: The populations are identical.
       H_A: $\mu_1 > \mu_2$
2. $\alpha = 0.05$
3. Reject the null hypothesis if $z \geq 1.645$; otherwise, accept H_0 or reserve judgment.
4. Arranging the data according to size, we get 55.35, 60.78, 62.63, 63.12, 63.76, 64.19, 66.51, 67.89, 67.95, 70.50, 70.78, 71.15, 71.72, 72.0, 75.16, 75.38, 75.91, 78.19, 78.60 and 86.45. Ranked in this order, the values of the first sample occupy ranks 6, 8, 9, 10, 11, 12, 14, 16, 19, and 20, so that $W_1 = 6 + 8 + 9 + 10 + 11 + 12 + 14 + 16 + 19 + 20 = 125$, $U_1 = 125 - \dfrac{10 \cdot 11}{2} = 70$, $\mu_{U_1} = \dfrac{10 \cdot 10}{2} = 50$, $\sigma_{U_1} = \sqrt{\dfrac{10 \cdot 10 \cdot 21}{12}} \approx 13.23$, and
$$z = \frac{70 - 50}{13.23} \approx 1.51 .$$
5. Since $z = 1.51$ is less than 1.645, the null hypothesis cannot be rejected; there is no real evidence to support the claim.

13.25 1. H_0: $\mu_1 = \mu_2 = \mu_3$
       H_A: $\mu_1, \mu_2,$ and μ_3 are not all equal
2. $\alpha = 0.05$
3. Reject the null hypothesis if $H \geq 5.991$ which is the value of χ^2 for $3 - 1 = 2$ degrees of freedom; otherwise accept it or reserve judgment.

4.   Arranging the data according to size, we get 164, 164, 166, 167, 168, 169, 169, 171, 173, 174, 176, 177, 178, 179, 181, 181, 183, and 189. Assigning the data, in this order, the ranks 1, 2, 3, ..., 18 we find that

$$R_1 = 14 + 1 + 15.5 + 11 + 9 + 13 + 2 = 65.5$$
$$R_2 = 6.5 + 12 + 3 + 8 + 15.5 + 18 = 63.0$$
$$R_3 = 10 + 4 + 5 + 17 + 6.5 = 42.5$$

and it follows that:

$$H = \frac{12}{n(n+1)} \cdot \sum_{i=1}^{k} \frac{R_i^2}{n_i} - 3(n+1) = \frac{12}{18 \cdot 19} \left(\frac{65.5^2}{7} + \frac{63.0^2}{6} + \frac{42.5^2}{5} \right) - 3 \cdot 19 = 0.39$$

5.   Since 0.39 is less than 5.991, the value of $\chi_{0.05}^2$ for 2 degrees for 2 degrees of freedom, we cannot reject the null hypothesis. There is no evidence at the 0.05 level of significance that the audience are not of equal size.

13.27   1. H_0: The populations are identical.

                H_A: The μ's are not all equal.

      2. $\alpha = 0.01$

      3.   Reject the null hypothesis if $H \geq 9.210$ which is the value of $\chi_{0.01}^2$ for $3 - 1 = 2$ degrees of freedom; otherwise, accept H_0 or reserve judgment.

      4.   Arranging the data according to size, we get 264, 276, 280, 280, 284, 284, 287, 287, 288, 289, 293, 293, 295, 296, 298, 299, 300, 303, 306, 309, 310, 311, 312, and 323. Ranked in this order from 1 through 24, the values of the first sample occupy ranks 2, 5.5, 9, 11.5, 17, 19, 20, and 23, those of the second sample occupy ranks 3.5, 5.5, 7.5, 11.5, 13, 15, 16, and 21, and those of the third sample occupy ranks 1, 3.5, 7.5, 10, 14, 18, 22, and 24. Therefore,

$$R_1 = 2 + 5.5 + 9 + 11.5 + 17 + 19 + 20 + 23 = 107,$$
$$R_2 = 3.5 + 5.5 + 7.5 + 11.5 + 13 + 15 + 16 + 21 = 93, \text{ and}$$
$$R_3 = 1 + 3.5 + 7.5 + 10 + 14 + 18 + 22 + 24 = 100. \text{ It follows that}$$

$$H = \frac{12}{24 \cdot 25} \left(\frac{107^2}{8} + \frac{93^2}{8} + \frac{100^2}{8} \right) - 3 \cdot 25 = 0.245.$$

      5.   Since $H = 0.245$ does not exceed 9.210, the null hypothesis cannot be rejected; the difference among the average weekly earnings are not significant.

13.29   1. H_0: Arrangement is random.

                H_A: Arrangement is not random.

      2. $\alpha = 0.05$

      3.   Reject the null hypothesis if $u \leq 8$ or $u \geq 19$, where 8 and 19 are the values of $u'_{0.025}$ and $u_{0.025}$ for $n_1 = 13$ red cards (R) and $n_2 = 12$ black cards (B), otherwise accept it or reserve judgment.

      4. $u = 25$ by inspection of data.

      5.   Since $u = 25$ is greater than 19, the critical value of $u_{0.025}$, the null hypothesis of randomness must be rejected.

13.31   1. H_0: The arrangement is random.

                H_A: The arrangement is not random.

      2. $\alpha = 0.05$

      3.   Reject the null hypothesis if $u \leq 7$ or $u \geq 7$; where 7 and 17 are the values of $u'_{0.025}$ and $u_{0.025}$ for $n_1 = 12$ and $n_2 = 10$ otherwise, accept H_0 or reserve judgment.

4. $u = 17$

5. Since $u = 17$ equals 17, the null hypothesis must be rejected; we conclude that the arrangement is not random. (There seems to be an alternating, or cyclical pattern.)

13.33 1. H_0: The arrangement is random.
H_A: The arrangement is not random.

2. $\alpha = 0.05$

3. Reject the null hypothesis if $z \leq -1.96$ or $z \geq 1.96$ otherwise, accept H_0 or reserve judgment.

4. $n_1 = 38$, $n_2 = 12$, and $u = 17$, so that $\mu_u = \dfrac{2 \cdot 38 \cdot 12}{50} + 1 = 19.24$

$\sigma_u = \sqrt{\dfrac{2 \cdot 38 \cdot 12(912 - 50)}{50^2 \cdot 49}} \approx 2.533$, and $z = \dfrac{17 - 19.24}{2.533} \approx -0.88$.

5. Since $z = -0.99$ falls between -1.96 and 1.96, the null hypothesis cannot be rejected; there is no real indication of any lack of randomness.

13.35 1. H_0: The arrangement is random.
H_A: The arrangement is not random.

2. $\alpha = 0.05$

3. Reject the null hypothesis if $z \geq -1.96$ or $z \geq 1.96$ otherwise, accept H_0 or reserve judgment.

4. $n_1 = 20$, $n_2 = 36$, and $u = 28$, so that $\mu_u = \dfrac{2 \cdot 20 \cdot 36}{56} + 1 \approx 26.71$,

$\sigma_u = \sqrt{\dfrac{2 \cdot 20 \cdot 36(1{,}440 - 56)}{56^2 \cdot 55}} \approx 3.40$, and $z = \dfrac{28 - 26.71}{3.40} \approx 0.38$.

5. Since $z = 0.38$ falls between -1.96 and 1.96, the null hypothesis cannot be rejected; we conclude that the signal may be regarded as random noise.

13.37 Practice exercise.

13.39 1. H_0: The arrangement is random.
H_A: The arrangement is not random.

2. $\alpha = 0.05$

3. Reject the null hypothesis if $z \leq -1.96$ or $z \geq 1.96$; otherwise, accept H_0 or reserve judgment.

4. The median is $\tilde{x} = \dfrac{65 + 67}{2} = 66$; and we get the following arrangement of values above and below the median: a a a a a a a a a a b b a a a a b b a a b b b a a b b b b b b b b a b b b b a b. As can be seen, $n_1 = 20$, $n_2 = 20$, and $u = 12$, so that $\mu_u = \dfrac{2 \cdot 20 \cdot 20}{40} + 1 \approx 21$,

$\sigma_u = \sqrt{\dfrac{2 \cdot 20 \cdot 20(800 - 40)}{40^2 \cdot 39}} \approx 3.12$ and $z = \dfrac{12 - 21}{3.12} \approx -2.88$.

5. Since $z \approx -2.88$ is less than -1.96, the null hypothesis must be rejected; there appears to be a trend, with the grades decreasing as the students take longer to finish the examination.

13.41 1. H_0: The arrangement is random.
H_A: The arrangement is not random.
2. $\alpha = 0.05$
3. Reject the null hypothesis if $z \le -1.96$ or $z \ge 1.96$; otherwise, accept H_0 or reserve judgment.
4. Arranging the data exactly as it appears in the exercise, we get the following values above, a, and below, b, the median value of 6.3:

$$
\begin{array}{cccccccccc}
b & b & b & b & a & b & b & b & a & b \\
a & b & b & a & b & b & a & b & b & b \\
b & b & a & b & b & a & b & a & a & b \\
b & a & a & b & a & a & a & b & a & a \\
a & a & a & a & a & b & a & a & a & a
\end{array}
$$

As can be seen, $n_1 = 25$, $n = 25$ and $u = 24$. Noting $\mu_u = \dfrac{2 \cdot 25 \cdot 25}{25 + 25} + 1 = 26$,

$$\sigma_u = \sqrt{\frac{2 \cdot 25 \cdot 25(1{,}250 - 50)}{50^2 \cdot 49}} \approx 3.50 \text{ and } z = \frac{24 - 26}{3.50} \approx -0.57.$$

5. Since $z = -0.57$ falls between -1.96 and 1.96, the null hypothesis cannot be rejected.

13.43 Practice exercise.

13.45 The ranks of the x's are 2, 10, 3, 1, 4, 8, 5, 12, 11, 6, 7, and 9; The ranks of the y's are 5, 7, 1, 4, 3, 12, 2, 11, 6, 10, 8, and 9; and the differences are -3, 3, 2, -3, 1, -4, 3, 1, 5, -4, -1, and 0.
Therefore, $\sum d^2 = (-3)^2 + 3^2 + 2^2 + (-3)^2 + 1^2 + (-4) + 3^2 + 1^2 + 5^2 + (-4)^2 + (-1)^2 + 0^2 = 100$
and $r_s = 1 - \dfrac{6 \cdot 100}{12(12^2 - 1)} \approx 0.65.$

13.47 The ranks of the x's are 8, 5, 2, 10, 4, 7, 9, 3, 6, and 1; the ranks of the y's are 8, 5, 3, 10, 4, 7, 9, 2, 6, and 1; and the differences are 0, 0, -1, 0, 0, 0, 0, 1, 0, and 0. Therefore,
$\sum d^2 = 0^2 + 0^2 + (-1)^2 + 0^2 + 0^2 + 0^2 + 0^2 + 1^2 + 0^2 + 0^2 = 2$ and $r_s = 1 - \dfrac{6 \cdot 2}{10(10^2 - 1)} \approx 0.99.$

13.49 The ranks of the x's are 4, 1, 7, 3, 5, 2, and 6; the ranks of the y's are 4, 1, 6, 3, 7, 2, and 5; and the differences are 0, 0, 1, 0, -2, 0, and 1. Therefore,
$\sum d^2 = 0^2 + 0^2 + 1^2 + 0^2 + (-2)^2 + 0^2 + 1^2 = 6$ and $r_s = 1 - \dfrac{6 \cdot 6}{7(7^2 - 1)} = 0.893.$ This is close to 0.885, the value of r obtained in Exercise 12.27.

13.51 The differences are 2, -1, 3, -5, -1, -2, -2, 1, 4, 4, -3, -1, 5, -5, and 1; so that
$\sum d^2 = 2^2 + (-1)^2 + 3^2 + (-5)^2 + (-1)^2 + (-2)^2 + (-2)^2 + 1^2 + 4^2 + 4^2 + (-3^2) + (-1)^2 + 5^2$
$+ (-5)^2 + 1^2 = 142$ and $r_s = 1 - \dfrac{6 \cdot 142}{15(15^2 - 1)} \approx 0.75.$

R.157 $n = 10$, $\sum x = 31.4$, $\sum x^2 = 99.88$, $\sum y = 27.6$, $\sum y^2 = 78.96$, and $\sum xy = 88.23$, so that

$S_{xx} = 99.88 - \dfrac{31.4^2}{10} = 1.284$, $S_{xy} = 88.23 - \dfrac{(31.4)(27.6)}{10} = 1.566$, $\bar{x} = 3.14$, and $\bar{y} = 2.76$.

Then $b = \dfrac{1.566}{1.284} \approx 1.22$, and $a = 2.76 - 1.22(3.14) \approx -1.07$ and the equation is

$\hat{y} = -1.07 + 1.22x$. For $x = 3.5$ the predicted grade point is $\hat{y} = -1.07 + 1.22(3.5) = 3.2$.

R.159 (a) Since $r = 0.33$ exceeds 0.304, which is the value of $r_{0.025}$ for $n = 42$, it is significant.

  (b) Since $r = 0.33$ falls between -0.393 and 0.393, where 0.393 is the value of $r_{0.005}$ for $n = 42$, it is not significant.

R.161 1. H_0 : The populations are identical.

  $H_A : \mu_1 \neq \mu_2$

  2. $\alpha = 0.05$

  3. Reject the null hypothesis if $U \leq 8$, which is the value of $U'_{0.05}$ for $n_1 = 7$ and $n_2 = 7$; otherwise, accept H_0 or reserve judgment.

  4. Arranging the data according to size, we get 10.0, 10.5, 10.8, 11.1, 11.3, 11.4, 11.5, 11.6, 11.7, 11.8, 11.9, 12.0, 12.1, and 12.2. Ranked in this order, from 1 through 14, the values of the first sample occupy ranks 5, 7, 9, 10, 11, 12, and 14, and those of the second sample occupy ranks 1, 2, 3, 4, 6, 8, and 13. Thus,

  $W_1 = 5 + 7 + 9 + 10 + 11 + 12 + 14 = 68$, $W_2 = 1 + 2 + 3 + 4 + 6 + 8 + 13 = 37$,

  $U_1 = 68 - \dfrac{7 \cdot 8}{2} = 40$, $U_2 = 37 - \dfrac{7 \cdot 8}{2} = 9$, and $U = 9$.

  5. Since $U = 9$ is greater than 8, the null hypothesis cannot be rejected; there is no real evidence that the two fishing lines are not equally strong.

R.163 1. $H_0 : \mu_1 = \mu_2$

  $H_A : \mu_1 < \mu_2$

  2. $\alpha = 0.05$

  3.′ The statistic is x, the number of plus signs.

  4.′ Replacing each positive difference with a plus sign and each negative difference with a minus sign, we get $---+---+----$, and $x = 2$. From Table I with $n = 12$ and $p = 0.50$, the probability of 2 or fewer successes is $0.003 + 0.016 = 0.019$.

  5.′ Since 0.019 is less than 0.05, the null hypothesis must be rejected; we conclude that conditions were such that golfers performed better on the first Sunday.

R.165 1. $H_0 : \rho = 0$

$H_A : \rho \neq 0$

2. $\alpha = 0.05$

3. Reject the null hypothesis if $r \leq -0.632$ or $r \geq 0.632$ where 0.632 is the value of $r_{0.025}$ for $n = 10$; otherwise, accept H_0 or reserve judgment.

4. $r = -0.65$ from Exercise R.164.

5. Since $r = -0.65$ is less than -0.632, the null hypothesis must be rejected; we conclude that there is a relationship between the amount of time a person spends reading books or magazines and the amount of time he or she spends watching television.

R.167 (a) Persons travel less when gasoline prices are high, and we can expect a negative correlation.

(b) Since exposure to the sun is a cause of skin cancer, we can expect a positive correlation.

(c) Since there is no relationship between shoe size and years of education, we can expect a zero correlation.

(d) Since scarcity leads to high values, we can expect a negative correlation.

(e) Since there is no relationship between blood pressure and eye color, we expect a zero correlation. Moreover, eye colors are not numeric.

(f) Since salaries are usually based on performance, we can expect a positive correlation.

R.169 Since the value which equals 149 has to be discarded, the sample size is only $n = 11$.

1. $H_0 : \tilde{\mu} = 149$

$H_A : \tilde{\mu} > 149$

2. $\alpha = 0.05$

3.' The statistic is x, the number of plus signs.

4.' Replacing each value greater than 149 with a plus sign and each value less than 149 with a minus sign, we get $+++--++++++$, and $x = 9$.

From Table I with $n = 11$ and $p = 0.50$, the probability of 9 or more successes is $0.027 + 0.005 = 0.032$.

5.' Since 0.032 is less than 0.05, the null hypothesis must be rejected; we conclude that the average attendance at such dances exceeds 149.

R.171 $n = 7$, $\sum x = 283$, $\sum x^2 = 11{,}815$, $\sum y = 280$, $\sum y^2 = 11{,}422$, and $\sum xy = 11{,}558$, so that $S_{xx} = 11{,}815 - \frac{283^2}{7} \approx 373.714$, $S_{xy} = 11{,}558 - \frac{(283)(280)}{7} = 238$, $\bar{x} = \frac{283}{7} \approx 40.429$, and $\bar{y} = \frac{280}{7} = 40$. Then $b = \frac{238}{373.714} \approx 0.637$, and $a = 40 - 0.637(40.429) \approx 14.247$ and the equation is $\hat{y} = 14.247 + 0.637x$. For $x = 40$ we get $\hat{y} = 14.247 + 0.637(40) = 39.727$ minutes.

R.173 1. H_0 : The populations are identical.

H_A : The μ's are not all equal.

2. $\alpha = 0.05$

3. Reject the null hypothesis if $H \geq 7.815$ which is the value of $\chi^2_{0.05}$ for $4 - 1 = 3$ degrees of freedom; otherwise, accept H_0 or reserve judgment.

4. Arranging the data according to size, we get 8, 9, 10, 12, 14, 15, 17, 19, 20, 21, 21, 23, 24, 25, 25, 25, 27, 28, 31, and 38. Ranked in this order from 1 through 20, the values of

the first sample occupy ranks 9, 12, 15, 19, and 20, those of the second sample occupy ranks 2, 4, 8, 10.5, and 13, and those of the third sample occupy ranks 1, 3, 5, 15, and 17, and those of the fourth sample occupy ranks 6, 7, 10.5, 15, and 18. Therefore,

$R_1 = 9 + 12 + 15 + 19 + 20 = 75.0$, $R_2 = 2 + 4 + 8 + 10.5 + 13 = 37.5$,

$R_3 = 1 + 3 + 5 + 15 + 17 = 41.0$, $R_4 = 6 + 7 + 10.5 + 15 + 18 = 56.5$, and

$$H = \frac{12}{20 \cdot 21}\left(\frac{75.0^2}{5} + \frac{37.5^2}{5} + \frac{41.0^2}{5} + \frac{56.5^2}{5}\right) - 3(21) \approx 5.03.$$

5.   Since $H = 5.03$ does not exceed 7.815, the null hypothesis cannot be rejected; there is no real evidence that the population means are not all equal.

R.175   Since one of the pairs, 18 and 18, must be discarded, the sample size is only $n = 23$.

1. $H_0 : \mu_1 = \mu_2$
     $H_A : \mu_1 \neq \mu_2$

2. $\alpha = 0.05$

3.   Reject the null hypothesis if $z \leq -1.96$ or $z \geq 1.96$; otherwise, accept H_0 or reserve judgment.

4.   Replacing each positive difference with a plus sign and each negative difference with a minus sign, we get $++++-++ +-+-+-+++-+++++-$ and $x = 17$. Therefore

$$z = \frac{17 - 23(0.50)}{\sqrt{23(0.50)(0.50)}} \approx 2.29.$$

5.   Since $z = 2.29$ exceeds 1.96, the null hypothesis must be rejected; we conclude that on the average there are not equally many burglaries in the two suburbs.

R.177   1. $H_0 :$ The arrangement is random.

     $H_A :$ The arrangement is not random.

2. $\alpha = 0.05$

3.   Reject the null hypothesis if $u \leq 6$ or $u \geq 16$, are the values of $u'_{0.025}$ and $u_{0.025}$ for $n_1 = 14$ and $n_2 = 8$; otherwise, accept H_0 or reserve judgment.

4. $u = 11$

5.   Since $u = 11$ falls between 6 and 16, the null hypothesis cannot be rejected; there is no real indication of any lack of randomness.

R.179   1. $H_0 :$ The populations are identical.

     $H_A : \mu_1 \neq \mu_2$

2. $\alpha = 0.05$

3.   Reject the null hypothesis if $U \leq 37$ which is the value of $U'_{0.05}$ for $n_1 = 12$ and $n_2 = 12$; otherwise, accept H_0 or reserve judgment.

4.   Since the values of the second sample occupy ranks 1, 2, 3, 5, 6.5, 6.5, 9, 10.5, 10.5, 15, 21, and 23, we get $W_2 = 1 + 2 + 3 + 5 + 6.5 + 6.5 + 9 + 10.5 + 10.5 + 15 + 21 + 23 = 113$,

and $U_2 = 113 - \dfrac{12 \cdot 13}{2} = 35$. From the preceding example, $U_1 = 109$ and it follows that

$U = 35$.

5.   Since $U = 35$ is less than 37, the null hypothesis must be rejected; we conclude that on the average the two insecticides are not equally effective.

R.181   1. $H_0 : \beta = 9.5$

           $H_A : \beta < 9.5$

     2. $\alpha = 0.01$

     3.   Reject the null hypothesis if $t \le -2.998$, where 2.998 is the value of $t_{0.01}$ for $9 - 2 = 7$ degrees of freedom; otherwise, accept H_0 or reserve judgment.

     4.   Find $S_{yy} = 486{,}017 - \dfrac{1{,}975^2}{9} \approx 52{,}614.23$ and $s_e = \sqrt{\dfrac{52{,}614.22 - \dfrac{(3{,}864.889)^2}{519.556}}{7}} \approx 58.39$

         and $t = \dfrac{7.439 - 9.5}{58.39} \sqrt{519.556} \approx -0.805$.

     5.   Since $t = -0.805$ is greater than -2.998, the null hypothesis cannot be rejected; there is no real evidence that the slope of the regression line is less than 9.5.

R.183   1. $H_0 : \rho = 0$

           $H_A : \rho \ne 0$

     2. $\alpha = 0.01$

     3.   Reject the null hypothesis if $z \le -2.575$ or $z \ge 2.575$; otherwise, accept H_0 or reserve judgment.

     4. $z = 0.34\sqrt{50 - 1} = 2.38$.

     5.   Since $z = 2.38$ falls between -2.575 and 2.575, the null hypothesis cannot be rejected; we conclude that value obtained for r_s is not significant.